T0261381

Luminos is the Open Access monograph publishing program
from UC Press. Luminos provides a framework for preserving and
reinvigorating monograph publishing for the future and increases
the reach and visibility of important scholarly work. Titles published
in the UC Press Luminos model are published with the same high
standards for selection, peer review, production, and marketing as
those in our traditional program. www.luminosoa.org

Hydrohumanities

Hydrohumanities

Water Discourse and Environmental Futures

———

Edited by

Kim De Wolff, Rina C. Faletti, and Ignacio López-Calvo

UNIVERSITY OF CALIFORNIA PRESS

University of California Press
Oakland, California

© 2022 by Kim De Wolff, Rina C. Faletti and Ignacio López-Calvo

This work is licensed under a Creative Commons CC BY license. To view a
copy of the license, visit http://creativecommons.org/licenses.

Library of Congress Cataloging-in-Publication Data

Suggested citation: De Wolff, K., Faletti, R. C. and López-Calvo, I. (eds.)
Hydrohumanities: Water Discourse and Environmental Futures. Oakland:
University of California Press, 2022. DOI: https://doi.org/10.1525/luminos.115

Names: De Wolff, Kim, editor. | Faletti, Rina C., editor. |
    López-Calvo, Ignacio, editor.
Title: Hydrohumanities : water discourse and environmental futures /
    edited by Kim De Wolff, Rina C. Faletti, and Ignacio López-Calvo.
Description: Oakland, California : University of California Press, [2022] |
    Includes bibliographical references and index.
Identifiers: LCCN 2021018718 (print) | LCCN 2021018719 (ebook) |
    ISBN 9780520380455 (paperback) | ISBN 9780520380462 (ebook)
Subjects: LCSH: Hydrology—Environmental aspects. |
    Water-supply—Environmental aspects. | Water-supply—Management.
Classification: LCC GB665.H92 2022 (print) | LCC GB665 (ebook) |
    DDC 333.91—dc23
LC record available at https://lccn.loc.gov/2021018718
LC ebook record available at https://lccn.loc.gov/2021018719

31 30 29 28 27 26 25 24 23 22
10 9 8 7 6 5 4 3 2 1

To lively waters in their many forms.
—Kim De Wolff

To my daughter, Xochi, and my husband, David; and in memory of my parents,
Peg and Jim, whose legacies flow within the currents of my work.
—Rina C. Faletti

To my friend Juanmi Vicente Amores.
—Ignacio López-Calvo

CONTENTS

LISTS OF FIGURES AND MAPS

FIGURES

MAPS

PREFACE

Ruth Mostern

Water is an exceptional subject for contemporary and interdisciplinary humanistic thinking. When faculty and graduate students affiliated with the UC Merced Center for the Humanities began discussing potential topics for a two-year seminar, we sought an area of focus that would exemplify the urgency of humanistic scholarship in troubled times. Moreover, we envisioned a concentration that would permit diverse participants to engage with the world around them while framing critical and enduring questions about aesthetics, value, and power. Based at a recently-founded campus with an interdisciplinary and community-serving mission, we were determined to engage with the broadest possible constituency. We wanted to host a humanities seminar that welcomed the arts and social sciences, involved colleagues from science and engineering, and extended off campus to embrace farmers, artists, activists, land managers, and other members of our rural community. Water was the perfect topic.

The importance of water as an area of inquiry begins with its ubiquity on and beneath the surface of the earth and in the cells of living beings. Water's periodic extremes of drought and flood have shaped social change, water travel has supported both trade and conquest, water has inspired art and music, and it has served as a site of exploration and play. Evoked in rituals like rites of passage, baptism, and ablution and in metaphors about "sea changes" and "ebbs and flows," water is symbolically and metaphorically significant as well. I was honored to preside over the seminar as Interim Director of the Center during the 2015–2016 academic year, and to participate as a Seminar Fellow during the 2016–2017 year.

We wanted to explore a wide breadth of topics about water and culture and to invite as many people as possible into that conversation. To do that, we hosted an eventful program of activities. An art exhibit about colonial depictions of water-scapes and wet bodies in West Africa, Hawai'i, and India launched a collaboration among an art historian, a historian, and the university library. Interdisciplinary art and science field trips on endangered seasonal wetlands at the foot of the Sierra Nevada mountains invited youth and adult participants to make art, watch birds, and perform science. One conference took agricultural land managers, National Park Service staff, farmers, curators, and artists to Yosemite National Park and the Merced County Courthouse Museum to talk about water's numerous social meanings in the highly managed California landscape. Finally, we offered space to UC Merced faculty and graduate students, and to visiting speakers, to present research: on the patron saints of Haitian fishermen, the history of California water rights, activist resistance to fracking, historical gardens in the Punjab, contempo-rary refugees at sea in the Mediterranean, and many more topics besides. In short, at a time when common ground is often elusive, the topic of water allowed us to share ideas about how to sustain our lives and communities.

Inspired by our collective thinking and conversation about widely diverse aspects of water, we focused upon a substance that supports human existence and all life on earth. Our geographical vantage point, in the center of California's ecologically and socially fragile Central Valley, attuned us to the unpredictable arrival of rainstorms every winter and the shock of wildfires every summer. These harsh seasonal rhythms, upon which well-being in the Central Valley and the Sierra foothills depends, characterize a region where everyone talks about water. The seminar allowed us to make meaning in conversation with the farmers, work-ers, scientists, and activists with whom we shared that space. When we gazed through the windows of the campus conference rooms where we held our events, we saw the waterscapes we were discussing: reservoirs, canals, and sluices; the ranchlands they irrigated; and the egrets, cattle, and bobcats that congregated around them.

In keeping with contemporary humanistic theory, the Water Seminar allowed us to enact a more-than-human humanities. Treating water as an active agent of social change and centering on water as the focus of our questions allowed us to talk about H_2O molecules, fungi, ships, and concrete along with people, with-out assuming that humans deserved primacy in our narratives. We blurred the imagined divide between nature and culture, and we explored how humans, trees, and soil mutually constitute one another. We learned how people transform water when they dam, pollute, and channel it, and when they sing about it and sail upon it, too. We also came to nuanced understandings of how water shapes culture: inundating the homes of the most vulnerable people, turning turbines for the ben-efit of state power, and irrigating farmland as sites of profit, food, and backbreak-ing labor as well as habitats for plants and animals. People make and move water on every scale, and they do so in the context of their aesthetics, values, power relations, and politics. Water in turn makes and moves people. These dynamics are multidirectional and perpetual. Relationships between water and people are never static.

We hosted the Water Seminar during a time of both planetary and academic peril. We are living in an epoch of climate disaster and in an age of retreat from the liberal arts as a shared public good. We launched the seminar during the final year of a historic California drought, and we completed it amidst floods, fires, and landslides. It is in this context that this volume of essays seeks to understand how humans subsist in watery landscapes together with other beings, and what courses of action humanists envision and propose to activate more just water futures. Water, unpredictably excessive in some places and conspicuously absent in others, will transform human and planetary futures ever more dramatically in the decades to come. I am confident that the humanities, and the essays in this book, will play an important role in shaping our collective sense of what we desire and what we can imagine in our relationships with water.

Introduction

Hydrohumanities

Kim De Wolff and Rina C. Faletti

In the twenty-first century, a new water discourse is emerging, carried by the humanities. It focuses on cultural changes, such as an emphasis on gender and race differentiations in water relations, ways water features in urban design, and decolonial analyses of water practices. It is deeply informed by new materialist and posthumanist attention to the active role of water in its multiple materialities. It is interdisciplinary, engaging with the geosciences as easily as with the arts in working toward transforming water futures. We call this emerging discourse surrounding water-human-power relationships the *hydrohumanities*.

Water in the modern era has been the domain of engineers, hydrologists, and economists. In the nineteenth and twentieth centuries, large-scale infrastructure projects dominated the water landscape and its discourse of expertise. Many discussions continue to privilege scientific and engineering studies centered on ecosystems management, or a governance and policy focus that attends to water rights and justice (Reuss 2004; Swyngedouw 2004). These existing discussions operate under the assumption that water—in the singular—is a resource to be managed or commodified, whether equitably or otherwise. The late twentieth century saw a shift towards an awareness of the environmental and social consequences of focusing on water as a commodity. This shift was largely instigated by widespread resistance to the corporate enclosure of formerly public waters, whether through the mass production of single-serving plastic bottles by multinational conglomerates such as Nestlé, or via the commodification of the very rain itself, with schemes like the World Bank–mandated privatization of all water in Bolivia (Barlow and Clarke 2002; Olivera and Lewis 2004). From drought to deluge, climate extremes are mobilizing humanities scholars to think about water with a new sense of urgency.

This book emerges from a two-year thematic focus on water and the humanities at the University of California, Merced, which included a biweekly seminar, numerous public events, and two conferences. Together, the contributors to this book demonstrate how interdisciplinary cultural approaches grounded in the humanities can transform water conversations that address intensified environmental crises, by promoting interchanges that are far more inclusive than those dominated by techno-economic and policy concerns. In turn, each of the nine chapters, along with this introduction, responds to a central question: how can humanities thinkers lead diverse scholars and publics into uncertain environmental futures through explorations of water?

WATER AND THE HUMANITIES

We define the humanities as approaches to studying human (and more-than-human) experiences with nuanced attention to culture and power, where questions of what is, has been, and could be are always ethical, political, and in process. While the humanities are more traditionally understood as a collection of disciplines concerned with the human condition (Goldberg 2014), our definition focuses instead on common concerns across these divisions, or what James Clifford calls "the greater humanities" (2013). Some contributors to this volume do identify as historians, philosophers, or anthropologists, and others are far more comfortable being interdisciplinary. Most important is a shared understanding that humanities approaches are not merely a matter of trading quantitative for interpretive methods, or of comparing methods between disciplines, but rather of insisting that *all* knowledges are situated (Haraway 1988). At the same time, emphasizing culture and power signals their inseparability: questions of meaning and value are inherently questions of politics, broadly defined.

From the coeditors' situated positions at the UC Merced Center for the Humanities where this volume was conceived, California, within the context of the American West, has served as an environmental hydrology case study par excellence. Our offices, overlooking the Vernal Pools Reserve, Lake Yosemite, and the Le Grand irrigation canal, all within the Merced River watershed, were windows into all kinds of ongoing water politics. These included the hydraulic modernism of the American Gold Rush inscribed in regional industrial water systems design, and ongoing wetland mitigation conservation projects meant to compensate for the development of the campus itself. We surveyed nearby reservoirs where boats lay grounded on drought-cracked lake beds, and then nervously updated our Twitter feeds the following rainy season as flooded watersheds threatened state infrastructures with collapse. Immersion in these powerful land-and-waterscapes was the incubator for the hydrohumanities.

A brief overview of the discourses relating to California and the Western United States exemplifies a central point that each chapter in this volume advances for the hydrohumanities: hydraulic environments embody social and political power, as do the knowledges that circulate about them. Three literatures have been especially influential in informing critical understandings of water and power in humanities scholarship: histories of water, especially as associated with environmental history; interdisciplinary water studies, most notably those that bridge natural sciences and the humanities; and contemporary scholarship that thinks *with* water, particularly with attention to its materiality. Outlining these areas in more detail below, we map a trajectory from disciplinary to interdisciplinary hydrohumanities water scholarship.

Among water historians, Donald Worster is the acknowledged principal for the United States: his 1985 book *Rivers of Empire: Water, Aridity, and the Growth of the American West*, an ecocultural history of industrial water systems development, led to his part in founding the subfield of environmental history, new in the 1970s. Worster's contribution was to reach beyond the United States to define regional water supply development as a centralized tool of hegemonic power wielding social and political control on a national scale. He grounded his water history in the theory of "hydraulic society" presented by Karl Wittfogel ([1957] 1967), a founding member of the Frankfurt School. Wittfogel's postwar study of communism in China had analyzed centralized control of water as bureaucratic ideological totalitarianism. Many subsequent scholars, while acknowledging Worster's formidable contribution, have objected to his sweeping application of Wittfogel's thesis to what they see as a dissimilar situation, challenging the attribution of a grand-scale overriding motivating factor—hegemonic imperialism—to California. Norris Hundley Jr. (2001), for example, preferred to look for diversity in local situations to uncover complexities of interplay between human values and waterscapes. Hundley has argued that water development scenarios are best understood—and critiqued—within their own specific contexts. This can reveal more accurately the ways political culture, policy-making, and cultural realities resonate within the positions and practices that imbue waterscapes with co-human agency.

The emergence of the subdiscipline of environmental history in the 1970s ran parallel with wide-ranging cultural dissemination of water politics in the popular press and in cultural production for the North American public at large. Perhaps most famously, Roman Polanski's 1974 film *Chinatown* dramatized the polemical politics of California's "water wars," spurred by the importation of water from distant watersheds into Southern California in the first few decades of the twentieth century. From the 1911 Los Angeles Aqueduct to the 1939 Colorado River Aqueduct and the range of federal and state irrigation projects of the 1950s to the '70s in California and Arizona, environmental justice responses ran deep. Citizen uprisings aimed at water injustices related to the funneling of the Owens River into the L.A. Aqueduct, for example, were to Southern California what San Francisco's Hetch Hetchy Aqueduct had been to Northern California, when the founding of the Sierra Club and related naturalist ideologies engaged protest against damming the Tuolumne River in Yosemite National Park. These issues live on the docket of water politics to this day. Environmental writer Marc Reisner's *Cadillac Desert: The American West and Its Disappearing Water* (1986) brought the politics of water into the public sphere in print, as *Chinatown* had done in film. Most recently, journalist Mark Arax contributed a 2019 best seller, *The Dreamt Land: Chasing Water and Dust across California*, a history-memoir-exposé expounding the inseparability of water and society. California-as-water-culture is a daily front-page feature, an ever-trending media theme, a matter of embedded public concern.

While California and the U.S. West anchored the inception of this volume at UC Merced, the Water Seminar participants, as well as the current volume's contributors, represented a broader balance of global water concerns in the humanities. An exemplary global-scale contribution to the hydrohumanities history discourse is the nine-volume series *A History of Water*, completed between 2005 and 2016 (Tvedt and Oestigaard, eds.). The volumes cover major themes in world water issues, with contributors from an array of disciplines and from every region of the world. The series focuses equally on the Global North and South, and on problems of water equity and access, into the future. One volume, for example, titled *Water and Food*, focuses twenty-two essays on Africa alone, to reveal the diversity and depth of water-related agriculture and food security throughout the continent's history (Tvedt and Oestigaard 2016). With the volumes released over the course of more than a decade, the series serves as a comprehensive chronicle of the world's major water issues at the beginning of the new millennium. Moreover, Terje Tvedt served as the first president of the International Water History Association (IWHA), founded in 2001 in parallel with the decision of the United Nations Educational, Scientific and Cultural Organization (UNESCO) to focus efforts on the roles historical scholarship could play in policy-making (Reuss 2004). Environmental history has long served as an international bridge between the technological and scientific aspects of water and water's investigation within the humanities.

In focusing on the role of humanities scholarship in leading toward sustainable water futures, *Hydrohumanities* is also indebted to a number of collaborative interdisciplinary volumes that establish the central place of culture in conversations about water. The product of a UNESCO project, *Water, Cultural Diversity and Global Environmental Change* (Johnston et al. 2011) anchors water as elemental in sustaining cultural and biological diversity in equal measure. The theoretical approach of *Hydrohumanities* more specifically builds on the groundbreaking collection *Thinking with Water* (Chen, MacLeod, and Neimanis 2013), as this work mobilizes diverse humanities perspectives to challenge the dominant Western assumption that water is a resource to be measured, managed, and sold. For Cecilia Chen and colleagues, alternative storyings and mappings of water bridge nature-culture and material-metaphor divides to connect meaning to ecological crisis. As Ingrid Stefanovic (2019) asks, nudging humanities water scholarship closer to publics and decision-making, "might a deeper, embodied vision of the wonder of water inspire more thoughtful policies?" *Hydrohumanities* builds on this shared foundation of relational thinking *with* waters in their many meaningful materialities, while attempting to push calls for alternative modes of attention within cultural theory toward new forms of hydraulic leadership.

Outstanding interdisciplinary volumes bridging arts, sciences, and beyond in discussions of water include *Rivers of the Anthropocene* (Kelly et al. 2017), which is primarily concerned with forming a transdisciplinary environmental research culture by exploring frameworks and methods for bridging natural sciences, social sciences, humanities, and policy. Taking a coordinating tack, this volume shows *what the humanities can do* in a time of crisis, and challenges the tendency to task the humanities with responding only after the natural sciences have identified a problem. We call on the humanities to lead.

Recently, there has been a proliferation of aquatic concepts and frameworks in the humanities. Instead of positioning water as an object *of* study, scholars are plunging into the material-conceptual depths to reimagine disciplines, forms, and approaches. There are disciplinary frameworks that push against traditions grounded by terrestrial bias: René ten Bos brings attention to Peter Sloterdijk's call for an "amphibious" anthropology equally at home in the water as on land, rather than privileging one element over the other (ten Bos 2009); Sugata Ray (2017) offers "hydroaesthetics" as methodological grounds for an ecological art history; and Laura Winkiel (2019) gathers leading literary ocean studies under a rubric of hydro-criticism. Where some scholars, such as Michelle Burnham (2019), mobilize oceans to rethink the story of the novel as a literary form, others are exploring how watery forms themselves overflow disciplinary boundaries. Irene Klaver follows the meander from its namesake river through classical design to a critique of efficiency culture (2014); while Stefan Helmreich surfs the waves of science past and present toward climate-changed futures (2014) and, with his colleague Caroline A. Jones (2018), looks at science, art, and culture through an "oceanic lens," with an eye for decolonial critique.

Humanities attention to oceans in particular has been generative of named bodies of water work. Most broadly, Elizabeth DeLoughrey (2017) describes a twenty-first-century humanistic "critical ocean studies," while others refer to the "blue humanities" (Gillis 2013), a term most often associated with the work of Steve Mentz and Stacy Alaimo. While blue humanities is sometimes deployed to be inclusive of those with freshwater foci, the concept has decidedly maritime origins, naming "an off-shore trajectory that places cultural history in an oceanic rather than terrestrial context" (Mentz 2018, 69). Indeed, there is a tendency for humanities water scholars to self-identify as ocean scholars or river scholars, separated by a salty/fresh divide. Historian of oceans Helen Rozwadowski (2010) traces this compartmentalization to the disparate origins of humanities river scholarship in environmental history and of humanities ocean scholarship in the history of science and technology. Of particular importance for this volume are the liminal spaces of aquatic environments and concepts, spaces the authors find especially productive for challenging existing boundaries, whether based on salinity or solidity. It is not enough to trade terrestrial bias for an aquatic one; these very categories, and the practices that constitute them, must be interrogated.

HYDROHUMANITIES

The hydrohumanities, then, emerge from much longer trajectories of studying water and power, while at the same time bridging water scholarship across fresh/salty divides. In some locales, the water prefix *hydro* is inseparable from power in a very practical sense. In British Columbia where large dam projects provide the bulk of electricity, for example, *hydro* is used as a synonym for electric power, as in having to pay one's hydro bill. As this colloquial shorthand reminds us, there is no power without water. Through the lens of hydropower, water is energy and force, but it can also be militaristic (DeLoughrey 2019) and hydrocolonial (Hofmeyr 2019). As we have begun to trace above, however, there have been notable shifts in how humanities scholars have conceptualized water-power relationships, from power *over* water, to water *as* power, to rethinking *with* water the very concept of power itself. Most importantly, hydrohumanities scholars see constellations of human-water-power relationships as irreducible to their component parts, none of which acts simply as a context for the others.

Discourses about water and power, however, have been predominantly focused on struggles for human power *over* water: who gets to own and control a limited resource with seemingly limitless economic potential. Many of these researchers' basic interests were built in part upon water law and policy, and on water rights history. These have ranged in focus from the exposure of social and environmental effects of water law and government policy (Pisani 1986) to comparisons of water governance structures around the world based on differences in their respective cultural histories (Dellapenna and Gupta 2009). Hundley's (2001) comprehensive California water history focuses on "human values and what human beings do to the waterscape" in Indigenous and industrial scenarios across the timeline (xviii). Water, in turn, is the object or the context, as DeLoughrey has noted of pre-1990s ocean scholarship, where "the ocean became a space for theorizing the materiality of history, yet it rarely figured as a material in itself" (2017, 33).

Others, however, focus on how water is itself powerful, not merely a substance to be fought over. Here, hydrohumanities scholarship has the potential to reinflate "flattened" ontologies where no entity (human or otherwise) is assumed to have more theoretical standing than another (as in actor-network theory or object-oriented ontology). Most crucially, water can help add dimension without limiting power to vertical hierarchies. Or, similarly, it can add depth where others have merely positioned "the sea as a stage for human history; a narrative of flat surfaces rather than immersions" (DeLoughrey 2017, 33). With Chandra Mukerji's conception of logistical power (2009), for example, water's properties are no mere context: water itself has agential capacities that can sometimes be shaped to assert human control over territories. "Water," she proclaims at the beginning of her chapter, "is an underestimated tool of power." Harnessed by the seventeenth-century French state engineering of the Canal du Midi, the resulting form of power is impersonal, enacted through "managing social relations by material means." Ruling with water, human control becomes almost imperceptibly enmeshed in waterscapes: "Logistical power worked silently outside social pressures, and in this sense, the canal seemed apolitical even as it initiated social change." Geographers, too, are rethinking power with water. Resisting the terrocentrism of the discipline, Christopher Bear and Jacob Bull (2011) emphasize that, in following the water, geographers must also "question the politics of moving through and with water" (2265). Taking this imperative more broadly, Philip Steinberg and Kimberley Peters (2015) outline an entire "wet ontology" drawing on voluminous oceanic depths to insist on dimension and nonlinear fluidity as a way to conceptualize a multidimensional spatiality of power and geopolitical order.

Conversations about water and power are often related to colonialism. This is especially paramount in settler-colonial societies where land-centric cultural and political systems have been violently imposed along with a decidedly Western conception of the very separation between land and water (Steinberg 2001). By contrast, a group of Indigenous scholars has outlined theories based on a "water view" *from* rather than *of* water (Risling Baldy and Yazzie 2018), recentering water in entangled processes of being, knowing, and responsibility in ongoing decolonial struggles. As Cutcha Risling Baldy and Melanie Yazzie explain, "our theoretical standpoint is one that foregrounds water view, (re)claiming knowledges not just for the people, but also for the water; not just looking at our relationship to water, but our accountability to water view" (2018, 2). Following Salish tradition, Lee Maracle rejects frameworks of Western ownership to remind us that "the water belongs to itself" (2017, 37).

As editors and settlers, we recognize the Wichita, Chinook, Clackamas, Wappo, and Yokuts peoples whose traditional lands we have inhabited and continue to occupy while working on this volume. Thus situated, *Hydrohumanities* is meant to be read alongside collections written and edited primarily by Indigenous water scholars and activists. Most notably, the special issue of *Decolonization* edited by Cutcha Risling Baldy and Melanie Yazzie (2018) provides a rigorous wealth of perspectives, while outlining how water has taken an unprecedented role as an "ideological and ontological centerpiece" within current waves of Indigenous resistance (8). The place of water in these struggles is perhaps most evident with the #NoDAPL movement's emblematic *mni wiconi*—"water is life"—a phrase that is at once prayer, ontology, and resistance. Water supports life, contains life, and is itself alive (Estes 2019, 13). For an even broader, globe-spanning set of perspectives, *Indigenous Message on Water* provides a multilingual anthology of water narratives, as knowledges representing the diversity of almost thirty Indigenous peoples (Sánchez Martínez and Quintanilla 2014). By contrast, *downstream: reimagining water* (Christian and Wong 2017) takes a more unifying tack, aiming to bridge European and Indigenous understandings of water to form an "intergenerational, interdisciplinary, culturally inclusive, participatory water ethic" (18). As the "decolonizing" of practices and discourses has become increasingly trendy in academia, it is crucial to position recognition as only a starting point in struggles that are fundamentally about the repatriation of Indigenous land/water/life (Tuck and Yang 2012). Scholars must actively stay mindful of ways in which we complicit in and continue to benefit from colonialism, even as we unpack questions about water, indigeneity, and justice.

MORE-THAN-ENVIRONMENTAL HUMANITIES

The once-nascent field of environmental humanities is now vast and growing. *Hydrohumanities* contributes to moving past conversations of definition (Oppermann and Iovino 2016) by establishing water as a key "arena" for the environmental humanities. In doing so, *Hydrohumanities* answers the call of Astrida Neimanis, Cecilia Åsberg, and Johan Hedrén (2015) for academic activism that incorporates "humanistic modes of inquiry into environmental problem-solving" (72). Enacting this vision, they explain, requires "a deeper and more open dialogue and integrated cooperation between the research community, policy-makers, society and ultimately private individuals" (74). This volume addresses the need to rethink scholarly approaches to environmental humanities through the lens of water. It represents both *tradition* and *transition* in the more-than-environmental humanities, where *tradition* draws upon interdisciplinary engagement to address the specific place of water in the environmental humanities, and where *transition* moves toward modeling a posthumanistic collaboration that enacts "a new configuration of knowledge" and "thick" citizen humanities practices (Nye et al. 2013; Åsberg 2014).

The environmental humanities must be a more-than-human humanities. From anthropogenic global climate change to synthetic plastics circulating in the blood of shellfish, the current situation makes it impossible to maintain any pretense that "the human condition" is separate from what has previously been bracketed off as "the environment." Following the now much-recited rejection of nature-culture dualisms, we approach waterscapes much as the editors of *Arts of Living on a Damaged Planet* do for landscapes, conceptualizing them as "overlaid arrangements of human and nonhuman living spaces" (Tsing et al. 2017, G1). Ocean scholars are already leading the way, learning *with* the sea toward an understanding that human histories have always been more-than-human (Mentz 2018, 69), and insisting that water bodies are transcorporeal bodies characterized by fluid material interchanges rather than by rigid boundaries (Alaimo 2012).

Theoretical interventions must be carried into practice. Though concerned with alternative conceptualizations, we build on field-defining efforts to *practice* environmental humanities by addressing water problems in the world. Here, Neimanis, Åsberg, and Hedrén (2015) stand out, promoting a naturalcultural ethics of encounter between and among scholars, institutions, and engaged publics working to cultivate a state of "'living well' with both human and more-than-human others in terms of responsivity" (2015, 82–84). With this conscious effort to enact praxis, we propose rethinking the "green field" (83) together with the blue field—water— as a whole. Niemanis et al. insist that this must be done "through the deployment of humanities modes of enquiry" (69). In fact, they conclude, "addressing these problems . . . is not possible *without* environmental humanities" (70, emphasis in original). Taken together, the essays that make up *Hydrohumanities* advocate for the work environmental humanities can do in the world and, specifically, what the environmental humanities can do with water.

BOOK ORGANIZATION

The chapters that follow are organized around three themes that characterize the hydrohumanities: agency of water, fluid identities, and cultural currencies. These themes emerged as common threads thickening over the course of the presentations and events that constituted the UC Merced Water Seminar. As these themes cross fluid forms, times, and spaces, they reveal their ability to show how humanities scholarship has world-changing potential. Each thematic section begins with a brief introduction that explores its respective concept and the associated intersections between its chapters in more detail. Across the volume as a whole, the chapters are organized in a trajectory from rigorously theoretical research toward explicit policy implications. Covering this continuum requires collaboration—with other humanities scholars and beyond—against persistent traditions of individualism.

Part I, *Agency of Water*, examines how water, under its own power, is harnessed by and ultimately confounds human desire and control. Many agencies are entangled in relationships with water, and the task of the hydrohumanities is to track them without losing sight of power. Chandra Mukerji anchors this section and the volume as a whole by foregrounding water in the emergence of a form of impersonal rule she calls logistical power. Focusing on the construction of the seventeenth-century French Canal du Midi, Mukerji meticulously details how water's agency is both harnessed by and comes into conflict with territorial governance. In demonstrating how hydraulic engineering becomes a tool of power, the lessons of the seventeenth century are relevant to present-day questions of how to borrow the power of water without destroying the natural world.

In chapter 2, Stephanie C. Kane takes up the concept of logistical power, reaching backward to the Ice Age in order to propel her narrative toward Anthropocene futures. Where Mukerji focuses on a modern canal, Kane turns to the potential channels forming in the wake of melting arctic ice fields, detailing how dramatic environmental changes offer new logistical power along some routes, while rendering others obsolete. Kane focuses on Winnipeg, Manitoba—a city located in the geographical center of North America—whose economic development looks toward warming trends to support a future role as a major global port. Given its current inland geographical location, this plan confounds the assumption that powerful port cities must be situated at strategic edges linking water routes and land markets. Kane details surprising intersections between geological and historical timespaces, as these set parameters for understanding the degradation of Arctic sea ice-as-solid-water. She emphasizes that more-than-human agency can be unintentional. Dwelling with seeming double contradictions—of an inland port and of the ocean in solid form—she demonstrates how the stakes of material-conceptual boundaries between land and sea emerge with and as conditions of possibility.

Much as Kane shows with her focus on ice-as-water, Irene J. Klaver in chapter 3 expands upon a multiplicity of waters that cannot be contained in a single elemental form. Building from the assertion that water is fundamentally relational, Klaver argues that water is "radical" in its refusal to be reduced to a singular hydro-logic of knowing and being. Detailing colonial water control projects shaping New York and New Orleans, Klaver shows how modern structures have attempted to make water into manageable, measurable entities—the delimited canals, dammed rivers, and determined routes detailed by Mukerji and Kane. In opposition to projects of domination and control, Klaver outlines an epistemology and ontology of meandering, urging scholars and engineers alike to give up rigid ways of thinking and living, to instead allow water—in all its relations—to be our guide.

Part II, *Fluid Identities*, further connects such commingled transformations of physical and conceptual waters to shifting urban and national identities. Ignacio López-Calvo and Hugo Alberto López Chavolla bridge book sections by studying the life-and-death significance of water for Latin American Indigenous communities, via South American literary fiction. First, attending to water's agency using the theoretical perspective of new materialism, they concentrate on the symbolic, cultural, magical, and salvational significance of mountain rivers for Peruvian Quechua communities in José María Arguedas's novel *Los ríos profundos* (*Deep Rivers*, 1958). Second, they contrast this worldview, from an ecocritical perspective, with the importance of sustainable waters for the Wayuu Indigenous group in the Guajira Peninsula of Colombia, as represented in Philip Potdevin's novel *Palabrero*. More-than-human interrelationships take primacy here, as Potdevin portrays the exponential rise in suicide rates among the Wayuu after international mining companies began to steal water resources from ancestral Wayuu lands.

Moving from Indigenous mountain rivers, Penelope K. Hardy's chapter 5 takes a dive deep into the fluid identity of the ocean. Hardy considers how historical developments in the methods and motivations for mapping the seafloor affected resulting Western definitions of ocean space and the uses to which these new identities were put. Exploring examples from the United States, Britain, Monaco, and Germany, Hardy shows that oceanic maps and charts were almost always constructed in the service of empire, so that the identities these maps assigned to the ocean itself reflected commercial and political as well as scientific concerns. During the nineteenth and twentieth centuries, both naval officers and naturalists increasingly attempted to chart greater depths, creating new Western scientific models of the ocean's three dimensions mediated by technologies of collection and measurement and by the motivations of those who wielded them as tools of territorial power. As a result, the ocean floor is mapped in terrestrial fashion right onto national and imperial frameworks of exploitation and control.

Following the course of aqueous national identities, Kale Bantigue Fajardo hones in on the city of Malolos, considered the birthplace of the Philippine Republic in 1898. His chapter 6 details how national imaginaries have broadened to include images of an "aquapelagic Malolos" of canoes, rivers, estuaries, coasts, and islands. Drawing on fieldwork, media analysis, and personal experience, Fajardo observes how cultural production, local tourism, and NGO activities have decentered church and mestizx architecture, drawing attention toward Malolos's relationships with water. Such a refocusing on water can instigate a precolonial and decolonial turn away from land-bound national monuments. Fajardo argues that the contemporary Philippine postcolonial nation-state, together with local and provincial governments, must move toward Manila Bay, that is, toward water and the seas, in order to better address climate change threats.

Part III, *Cultural Currencies*, presents both top-down and bottom-up challenges to the technical and economic logics that dominate public water conversations and influence policy-making. Here, we return, via culture, to conversations about policy in a narrower sense. In chapter 7, Rina C. Faletti connects an iconic 1970s scientific photograph representing land subsidence to present-day water policy by examining the politics of agricultural photography. In 1977, surrounded by vineyards in California's Central Valley, a prominent USGS hydrogeologist posed next to a power pole, marked high above his head with the years of previous decades, to show where the land surface lay in the past and to emphasize how dramatically it had subsided. With this photograph he staged a human-scale visualization of vast groundwater resources disappearing with unchecked industrial pumping. Situating the photograph within a broader history of California agricultural documentary photography between the 1920s and the 1980s, Faletti reveals a parallel subtext of societal conflicts caused by industrial water systems. California's water, agri-, and petro- cultures rose as the land surface, and the social substrata that labored to work on and in it, invisibly declined. Issues raised by photographers have fed policy interests into the present. In 2014, California's Sustainable Groundwater Management Act set forth a plan for statewide groundwater stewardship into the coming century. Water-inspired images like those under Faletti's analysis reveal long-term effects of both scientific and social documentary photographers, whose visual calls for action work toward remedying otherwise invisible problems of environmental and social justice pertaining to enmeshed relationships among land, water, and work.

In chapter 8, James L. Wescoat Jr. and Abubakr Muhammad urge scholars—and policy makers—to incorporate into water management the cultural concept of the Indus basin as a garden. Since the second half of the twentieth century, water managers have described the Indus River basin as a food machine, an irrigation system, and a water-energy-food nexus, while social scientists continue to emphasize issues of governance, power dynamics, and political economy. Turning to future problem-solving for postcolonial development of the Indus basin, Wescoat and Muhammed detail continuing references to Edenic ideals in the region, while documenting the eclipse of Eden by techo-metaphors: machine, system, nexus. The explicit exclusion of garden irrigation from the 1960 Indus Waters Treaty prompts a thought experiment: what difference might it make to people and places of the Indus region to return to the idea of the Indus basin as a garden? The answer from Wescoat and Muhammed is a call for a culturally-based model of water resource management.

Finally, Veronica Strang concludes the volume by opening humanities water policy conversations situated on a global scale, recounting her participation in the process of defining *culture* in the United Nations' 2017 Guiding Principles for Water. Policy debates about water's *value* tend to focus on conceptualizations of water that can be measured quantitatively, with water's cultural value defined as a discrete area of use, vaguely associated with spiritual meanings or cultural heritage. Traditionally resistant to quantification, cultural values of water are placed on the periphery of decision-making about water use and management, and are often ignored altogether. In counterpoint, Indigenous communities propose integrated views of water's value, offering the critique, often in collaboration with social sciences and humanities scholars, that the notion of culture as a domain apart from the real world is unproductive for water policy futures. In response, experts, including Strang, worked to define more inclusive categories for value, organizing the UN Principles into "economic values," "environmental values," and "cultural and spiritual values." Revising customary thinking in this way confronted several challenges: how to articulate "cultural values" for water that could be integrated into local, regional, and global decision-making; how to introduce a unifying theoretical basis to counter categorical divisions that deny water-as-culture; and how to demonstrate that cultural values of water are central not only to the environment, but to economic activities around the globe.

Together, these contributions call on humanities scholars to craft new stories of power and values by pursuing alternative mappings deepened by a multiplicity of views. This collection, as a whole, is meant to guide scholarly and public discourse in a current era that demands more creative, and more relevant, reimaginings of environmental issues, with approaches that can actualize more just water futures.

REFERENCES

Alaimo, Stacy. 2012. "States of Suspension: Transcorporeality at Sea." *Interdisciplinary Studies in Literature and Environment* 19 (3): 476–93.

Arax, Mark. 2019. *The Dreamt Land: Chasing Water and Dust across California*. New York: Alfred A. Knopf.

Åsberg, Cecilia. 2014. "Resilience Is Cyborg: Feminist Clues to a Post-Disciplinary Environmental Humanities of Critique and Creativity." *Resilience: Journal of Environmental Humanities* 1: 5–7.

Barlow, Maude, and Tony Clarke. 2002. *Blue Gold: The Fight to Stop the Corporate Theft of the World's Water*. New York: The New Press.

Bear, Christopher, and Jacob Bull. 2011. "Water Matters: Agency, Flows, and Frictions." *Environment and Planning A* 43 (10): 2261–66.

Burnham, Michelle. 2019. *Transoceanic America*. Oxford: Oxford University Press.

Chen, Cecilia, Janine MacLeod, and Astrida Neimanis, eds. 2013. *Thinking with Water*. Montreal: McGill-Queen's University Press.

Christian, Dorothy, and Ruth Wong, eds. 2017. *downstream: reimagining water*. Waterloo: Wilfrid Laurier University Press.

Clifford, James. 2013. "The Greater Humanities." *Occasion: Interdisciplinary Studies in the Humanities* 6 (October 1). http://arcade.stanford.edu/occasion/greater-humanities.

Dellapenna, Joseph W., and Joyeeta Gupta. 2009. *The Evolution of Law and Politics of Water*. New York: Springer.

DeLoughrey, Elizabeth. 2017. "Submarine Futures of the Anthropocene." *Comparative Literature* 69 (1): 32–44.

———. 2019. "Toward a Critical Ocean Studies for the Anthropocene." *English Language Notes* 57 (1): 21–36.

Estes, Nick. 2019. *Our History Is the Future: Standing Rock vs. the Dakota Access Pipeline and the Long Tradition of Indigenous Resistance*. London: Verso.

Gillis, John R. 2013. "The Blue Humanities." *The Humanities* 34 (3). www.neh.gov/humanities/2013/mayjune/feature/the-blue-humanities.

Goldberg, David Theo. 2014. "The Afterlife of the Humanities." University of California Humanities Research Institute. April. https://humafterlife.uchri.org.

Haraway, Donna. 1988. "Situated Knowledges: The Science Question in Feminism and the Privilege of Partial Perspective." *Feminist Studies* 14 (3): 575–99.

Helmreich, Stefan. 2014. "Waves: An Anthropology of Scientific Things." Transcript of Lewis Henry Morgan Lecture, Department of Anthropology, University of Rochester, October 22, 2014. *HAU: Journal of Ethnographic Theory* 4 (3): 265–84.

Helmreich, Stefan, and Caroline A. Jones. 2018. "Science/Art/Culture through an Oceanic Lens." *Annual Review of Anthropology* 47: 97–115.

Hofmeyr, Isabel. 2019. "Provisional Notes on Hydrocolonialism." *English Language Notes* 57 (1): 11–20.

Hundley, Norris, Jr. 2001. *The Great Thirst: Californians and Water—a History*. Revised ed. Berkeley: University of California Press.

Johnston, Barbara Rose, Lisa Hiwasaki, Irene J. Klaver, Ameyali Ramos Castillo, and Veronica Strang, eds. 2011. *Water, Cultural Diversity and Global Environmental Change: Emerging Trends, Sustainable Futures?* New York: Springer.

Kelly, Jason M., Philip Scarpino, Helen Berry, James Syvitski, and Michel Meybeck, eds. 2018. *Rivers of the Anthropocene*. Oakland: University of California Press.

Klaver, Irene J. 2014. "Meander(ing) Multiplicity." In *Mediterranean Mosaic: Water Scarcity, Security and Democracy*, edited by Gail Holst-Warhaft, Tammo Steenhuis, and Francesca de Châtel, 36–46. Ithaca, NY: Atkinson Center for a Sustainable Future, Cornell University; Athens: Global Water Partnership–Mediterranean.

Maracle, Lee. 2017. "Water." In *downstream: reimagining water,* edited by Dorothy Christian and Ruth Wong, 33–38. Waterloo: Wilfrid Laurier University Press.

Mentz, Steve. 2018. "Blue Humanities." In *Posthuman Glossary*, edited by Rosi Braidotti and Maria Hlavajova, 69–72. London: Bloomsbury.

Mukerji, Chandra. 2009. *Impossible Engineering: Technology and Territoriality on the Canal du Midi*. Princeton: Princeton University Press.

Neimanis, Astrida, Cecilia Åsberg, and Johan Hedrén. 2015. "Four Problems, Four Directions for Environmental Humanities: Toward Critical Posthumanities for the Anthropocene." *Ethics and the Environment* 20 (1): 67–97.

Nye, David, Linda Rugg, James Fleming, and Robert Emmett. 2013. *Emergence of the Environmental Humanities*. MISTRA. www.mistra.org/en/mistra/application-calls /completed-application-calls/environmental- humanities.html.

Olivera, Oscar, and Tom Lewis. 2004. ¡*Cochabamba! Water War in Bolivia*. Boston: South End Press.

Oppermann, Serpil, and Serenella Iovino, eds. 2016. *Environmental Humanities: Voices from the Anthropocene*. London: Rowman and Littlefield.

Pisani, Donald J. 1986. "Irrigation, Water Rights, and the Betrayal of Indian Allotment." *Environmental Review* 10 (3): 157–76.

Polanski, Roman, dir. 1974. *Chinatown* (motion picture). Hollywood: Paramount Pictures Corporation.

Ray, Sugata. 2017. "Hydroaesthetics in the Little Ice Age: Theology, Artistic Cultures and Environmental Transformation in Early Modern Braj, c. 1560–70." *South Asia: Journal of South Asian Studies* 40 (1): 1–23.

Reisner, Marc. 1986. *Cadillac Desert: The American West and Its Disappearing Water*. New York: Penguin Books.

Reuss, Martin. 2004. "Historians, Historical Analysis, and International Water Politics." *The Public Historian* 26 (1): 65–80.

Risling Baldy, Cutcha. and Melanie Yazzie. 2018. "Indigenous Peoples and the Politics of Water." *Decolonization: Indigeneity, Education & Society* 7 (1): 1–18.

Rozwadowski, Helen. 2010. "Oceans: Fusing the History of Science and Technology with Environmental History." In *A Companion to American Environmental History*, edited by Douglas Cazaux Sackman, 442–61. Malden, MA: Wiley-Blackwell.

Sánchez Martínez, Juan Guillermo, and Felipe Quetzalcoatl Quintanilla, eds. 2014. *Indigenous Message on Water*. London, Ontario: Indigenous World Forum on Water and Peace.

Stefanovic, Ingrid, ed. 2019. *The Wonder of Water: Lived Experience, Policy, and Practice*. Toronto: University of Toronto Press.

Steinberg, Philip E. 2001. *Social Construction of the Ocean*. Cambridge: Cambridge University Press.

Steinberg, Philip, and Kimberley Peters. 2015. "Wet Ontologies, Fluid Spaces: Giving Depth to Volume through Oceanic Thinking." *Environment and Planning D: Society and Space* 33: 247–64.

Swyngedouw, Erik. 2004. *Social Power and the Urbanization of Water: Flows of Power*. Oxford: Oxford University Press.

ten Bos, René. 2009. "Towards an Amphibious Anthropology: Water and Peter Sloterdijk." *Environment and Planning D: Society and Space*. 27 (1): 73–86.

Tsing, Anna, Heather Swanson, Elaine Gan, and Nils Bubandt, eds. 2017. *Arts of Living on a Damaged Planet*. Minneapolis: University of Minnesota Press.

Tuck, Eve, and K. Wayne Yang. 2012. "Decolonization Is Not a Metaphor." *Decolonization: Indigeneity, Education & Society* 1 (1): 1–40.

Tvedt, Terje, and Terje Oestigaard, eds. 2016. *A History of Water*, series 3, vol. 3, *Water and Food*. London: I.B.Tauris.

Winkiel, Laura. 2019. "Introduction." *English Language Notes* 57 (1): 1–10.

Wittfogel, Karl August. (1957) 1967. *Oriental Despotism: A Comparative Study of Total Power*. New Haven: Yale University Press.

Worster, Donald. 1985. *Rivers of Empire: Water, Aridity and the Growth of the American West*. Oxford: Oxford University Press.

PART I

Agency of Water

TO WRITE OF WATER'S AGENCIES is to incorporate the physical properties of water into scholarship beyond the domain of the natural sciences, and in doing so, to decenter the human in the humanities. As such, attending to material agency is a defining characteristic of broader "new material," "nonhuman," and "posthuman" turns (Bennett 2010; Braidotti 2016; Coole and Frost 2010; Grusin 2015). Agency in these traditions has generally been uncoupled from intention, with the very distinction between animate and inanimate subject to critique (Chen 2012). For scholars grappling with how more-than-human entities like water might have "desires" or "needs," it might make sense to simply ask how water might "come to matter" (De Wolff 2018). How does water not merely reflect human culture, but itself actively make a difference? Each of the three essays in this section emphasizes a different facet of water's agency, together demonstrating what water can do—for itself and for the hydrohumanities.

As accounts of materiality in the humanities have proliferated over the past few decades, there has been a tendency for discussions of materiality to become divorced from power. Chandra Mukerji's work stands apart. In her chapter 1, the material agency of water and the emergence of new forms of state power are inseparable. Water is a powerful substance whose properties—of motion, velocity, and change—can be harnessed by humans through hydraulic engineering. At the same time, water is a "trickster" that, "acting independent of human will," proves nearly impossible to govern. For Mukerji, water's agential autonomy is embodied by resistance: even as water becomes an agent of the state, it still "does what it wants," flowing over, under, and around human design. These tensions between water and human agency are a reminder of the limits of human control.

They are a reminder of the dangers of massive environmental interventions, where water may be an important—and impersonal—source of social change, but remains "an unruly material that defies easy control."

As the concept of logistical power carries through Stephanie C. Kane's chapter 2, water not only asserts its own agency, but also teaches us about the unintentionality of human agency. The story that unfolds on the slippery terrain of melting sea ice is an elemental reminder that liquid is not water's only state. As the icy territories of Arctic Indigenous peoples dissolve into warming seas, they open new channels of exploitation. Here, water in solid form slips between the dualistic water-land ontology grounding international legal rights frameworks. Kane challenges us to envision "a drama in which both humans and ice/water have agential powers that form and act materially and culturally on the stages of earth's amphibious crust." To understand such intertwined geo-cultural transformations requires expanding notions of human agency: though climate change is anthropogenic, humans have caused it unintentionally. The move toward a more "equitable" division of human-water agencies can help us to productively approach environmental crises in new ways: "shocking stories of climate change . . . can be told otherwise, and perhaps better questions can be asked."

While it is tempting to talk about *the* agency of water, or reduce its powers to gravitational forces, in chapter 3 Irene J. Klaver insists on the agency of water in all its multiplicities, where "water is relational." For Klaver, water is inherently "radical," because of its irreducibility to a singular root form, such as H_2O. Despite modern attempts to limit water into submission, water resists linear simplifications: "It challenges clear-cut divisions and oppositions, undermines categorizations, messes up lines of separation, laughs at institutions, builds and resists infrastructures. It leaks, overflows, erodes, spreads, disappears, dilutes, and pollutes." In doing so, water thwarts the ordered structures—both material and conceptual—of modern management. Klaver urges hydrohumanists instead toward slow, "meandering" relationships *with* water, relationships instigated by water itself: "Stressed to its limit, water demands radical change . . . in our thinking and doing."

Though each chapter emphasizes different implications for understanding water's agencies, collectively they insist that water cannot simply be conceptualized—or treated—as a passive substance to be exploited. These chapters point to at least three ways that hydrohumanities scholars can take the lead: First, by insisting that hydraulic infrastructures—whether canals, ports, shipping channels, levees, or land claimed from the sea—are infrastructures of social and political power. And so, too, are knowledges that make hydraulic infrastructures and that are made about them. As Jessica Budds (2009) has documented, this holds even with seemingly objective environmental knowledge: the rationality of water is the product of the scientific practices that render it measurable, not an inherent quality of water itself. Second, by resisting this deradicalization of water to an exploitable singularity, we can see instead how water "teaches us relationality, it teaches us to change: to live *with* it, instead of controlling it." As Klaver suggests, it is multiplicity that is an intrinsic property of water. A plurality of waters reduced to a singular, homogenous water can be traced to modern infrastructural ideals and colonial practices (Linton 2010; Walsh 2018). Third, by following water's lead—by making water and ice allies—we can rethink human-water agencies. As Kane asserts, humanists may have an advantage in understanding "significant unintentional drivers" of ecological crises. Though scholarly hope may lie in thoughtful human action, we cannot, she warns, "be so arrogant as to assume that the ability of humans to intend is sufficient." Through thinking and acting *with* water, the hydrohumanities can help define "environmental" problems and shape responses in their many entanglements.

REFERENCES

Bennett, Jane. 2010. *Vibrant Matter: A Political Ecology of Things*. Durham, NC: Duke University Press.

Braidotti, Rosi. 2016. *The Posthuman*. Cambridge: Polity Press.

Budds, Jessica. 2009. "Contested H$_2$O: Science, Policy and Politics in Water Resources Management in Chile." *Geoforum*, 40 (3): 418–30.

Chen, Mel Y. 2012. *Animacies: Biopolitics, Racial Mattering and Queer Affect*. Durham, NC: Duke University Press.

Coole, Diana, and Samantha Frost, eds. 2010. *New Materialisms: Ontology, Agency and Politics*. Durham, NC: Duke University Press.

De Wolff, Kim. 2018. "Materiality." Theorizing the Contemporary, *Fieldsights*, March 29, 2018. https://culanth.org/fieldsights/materiality.

Grusin, Richard, ed. 2015. *The Nonhuman Turn*. Minneapolis: University of Minnesota Press.

Linton, Jamie. 2010. *What Is Water? The History of a Modern Abstraction*. Vancouver: UBC Press.

Walsh, Casey. 2018. *Virtuous Waters*. Oakland: University of California Press.

The Agency of Water
and the Canal du Midi

Chandra Mukerji

Water is an underestimated tool of power. Its useful properties—what I would call agential properties—are so crucial to human life that hydraulic infrastructures can shape social relations and mediate power (Cronon 1992, 207–52; Lansing 1991). Water is used to power mills, carry boats, nourish animals, irrigate fields, take away refuse, and wash laundry. Water is not only necessary for the nourishment of living things, but it acts independently of human will in ways that can affect social possibilities. It flows and floods, evading capture and overcoming boundaries by flowing over and around them; or it can collect in low-lying areas, be tapped with wells, or disappear into the sand. Water's physical properties can be made to serve human communities through engineering, enhancing the agency of individuals or the powers of political regimes, but water can also be a trickster, defying or eluding human control. Both the powers of water and the difficulties in governing them are the reasons why hydraulic engineering can be a tool of power, or, more specifically, a legitimating means of impersonal rule.

The importance of water for public administration was made obvious in eventeenth-century France when the state authorized the construction of a canal through Languedoc—what would come to be called the Canal du Midi. The king, Louis XIV, indemnified land for the project, taking it away from local nobles, and the canal itself shifted the form of life around it, eroding traditional relations of power. It changed transport, manufacture, the location of mills, the movement of mail, and what people did with their laundry, in mundane ways altering the lives of people through the impersonal exercise of territorial governance.

In the roughly twenty years between 1663 and 1684, this navigational canal of 240 kilometers length and 50 locks was cut across Languedoc just north of the Pyrenees (Bergasse 1982–86; Mukerji 2009; Maistre 1998; Rolt 1973). The project was technically impossible according to the formal knowledge of hydraulics in the period, and so the waterway was a display of technical finesse so extraordinary that it was described after its completion as a "wonder of the world." The Canal du Midi was indeed surprising as it carried large boats across broad stretches of the region's arid landscape, floating on water that circulated around mountains and plains in ways water did not normally flow. In its own period, it was often called the *Canal des Deux Mers*, or the canal linking the two seas, because it crossed the continental divide, connecting the Garonne River near Toulouse (which was navigable to the Atlantic Ocean) to the Mediterranean Sea at a new port of Sète. It was a model of territorial governance and impersonal rule that stood for the king and his administration, but far from the person of the monarch.

The Canal du Midi was the most ambitious and expensive engineering project undertaken during the reign of Louis XIV, and was celebrated as evidence of the Sun King's ability to revive France's ancient engineering heritage and restore the glory of Gaul. It fit culturally with the king's ambitions for empire and propaganda efforts to equate France with Rome.

Material politics made sense in this period as Humanists revealed the greatness of Rome through things. Books and ruins were material remains that demonstrated how well the ancient empire deployed logistical powers for political ends. The French administration under Louis XIV's minister, Jean-Baptiste Colbert, studied these practices and tried to imitate them (Blair 2020; Miller 2000; Gébara and Michel 2002). Hydraulics had a central place in the cultural genealogy he was helping to draw from ancient Rome to modern France. Water was central to seventeenth-century understandings of Rome and to Colbert's policies of using engineering to empower the state at the Canal du Midi (Adkins and Adkins 1994; Long 2001).

Cutting a canal through Languedoc was also part of a broader political program to empower the monarchy using tools of impersonal rule. In Languedoc, turning to logistical politics was particularly important because the power of the northern monarch had become attenuated in that region during the Wars of Religion. Obedience to the sovereign was politically required, but not automatic for dissident nobles who routinely ignored or evaded the king's commands. It was this disobedience that spurred Colbert to seek alternative ways to enhance the power of the state, turning to material methods of territorial governance and legal practice that depersonalized relations to land and law (Mukerji 2007a). The resulting system has ironically been called state absolutism, or an enhanced form of personal rule. But Louis XIV's personal will was made more effective in this period only because the administration undercut the patrimonial order with impersonal forms of governance.[1] State agents avoided, as much as possible, personal confrontations that the king's rivals could resist, and used legal precedents and infrastructural projects to transform a weak monarchy into state absolutism.

In this period of French patrimonial politics, power was supposed to circulate through social networks, not derive from the manipulation of landscapes, papers, and libraries. Royal policies were meant to be expressions of God's will, not the product of natural knowledge and earthly practices. As Kettering (1986) has shown, French patrimonial power was exercised through social networks cemented by the circulation of information and favors. And as Beik (1985) has argued, noble families enjoyed a high degree of autonomy within their regions, avoiding constraints by the state and king. The French monarchy was chronically weak as a result, and this became clear during the Fronde in Louis XIV's youth when powerful nobles and members of the bourgeoisie rose up against the government. Even though the *Frondeurs* were finally defeated, their strength had demonstrated the contingent nature of royal power in France. Thus, when Louis XIV took the throne, he stepped into a position of political vulnerability. It was Colbert's job to change this, which he did systematically and effectively, using techniques of impersonal rule.

Jacob Soll (2009) has identified one method Colbert used: the collection and analysis of historical legal documents. Soll argues that the clergy was a problem for Louis XIV as well as the nobility. Louis XIV did not have personal means to subordinate the French church to the crown, so Colbert established a legal basis, following the Roman practice of using legal archives to discipline the people exercising legal powers. France had been administered for much of the seventeenth century by clerics. Two long regencies during the minority of young kings, including Louis XIV, had put Cardinals Richelieu and Mazarin in charge of the French state, increasing the authority of the church over French politics and normalizing this through law. Colbert strove to rectify this situation not by confronting the clergy, but rather by undercutting them, using a form of impersonal power. He collected legal documents and employed a librarian to study them to establish through precedent the ultimate authority of French kings since Charlemagne over politics in France. Books and archives became administrative weapons of political warfare just as territorial engineering would, too.

This story of monarchical weakness and the turn to legal documentation to subordinate the clergy to the king was paralleled in Languedoc with construction of the Canal du Midi and the subordination of the regional nobility. The king's personal authority had become attenuated in the region, but the monarch still had rights to indemnify land and an obligation to act as steward of his kingdom, so he had the power to authorize construction of an infrastructure project that effectively reduced noble control over local land. Water and stone, like papers and books, were surprising but effective tools of government, and used in Languedoc for the subordination of king's rivals to the crown.

Languedoc was a region where the nobility and peasantry alike had become particularly independent of the crown. Languedoc had been bloodied by the wars of religion. Towns were taken and retaken by Catholics and Huguenots, as families were displaced, buildings razed, fields burned, and lives destroyed. Royal power had been used to crush Huguenot uprisings, creating a rift between the Catholic monarchy and the Huguenot nobles that dominated much of the region. Peace had come, but many members of the local elite still saw no reason to subordinate themselves to a king who stood for a religion they understood as corrupt.

Elites had formal obligations to the monarch, but they were prone to ignore edicts from Paris (Holt 2001). Meanwhile, peasants used what James Scott has called the "weapons of the weak" (1985) to keep representatives of the king at bay, killing some tax collectors and threatening others (Froidour 1899; Mukerji 2009).

It was in this dissident, violent, and tumultuous region that a salt tax collector, Pierre-Paul Riquet, proposed to build a navigational canal to link the Mediterranean Sea to the Garonne River and the Atlantic Ocean. It was the perfect opportunity for Colbert to shake up local life by reworking the infrastructure. Riquet seemed an appropriate person for the project, too, because, as a tax farmer, he was already a contractor, doing dirty work for the state. He farmed the *gabelle* or salt tax, the most hated tax, and effectively brought in revenue in spite of local opposition. Riquet also proposed to build a canal wide enough to carry warships, not only explicitly allowing the French navy to move between the two seas without using the dangerous passage by Gibraltar, but also implicitly bringing the police power of the state to Languedoc in a permanent way.

Elites in this querulous corner of France opposed the project, assuming that their positions would be threatened and their land control compromised. But the project was as appealing to the king and Colbert as it was distasteful to locals. The minister had found a way to erode the habits of local life that sustained noble autonomy from the central government, giving the king new ways to impose his will on unruly subjects (Adgé 1992, 202–3; Mukerji 2009).

Riquet may have been a good salt tax collector, but he was no engineer. Still, he could imagine cutting a navigational canal across Languedoc because he traveled throughout the region on tax business and encountered local evidence of hydraulic sophistication. In the mountains and plains of Languedoc with dry summers and wet winters, water could be abundant or absent, so it was often captured at sources and rivers and channeled to where it was needed. Languedoc was riddled with ditches and diversionary channels that carried water towards towns, fields, and mills. A navigational canal seemed just a larger version of familiar cultural forms.

Languedoc was rich in hydraulic knowledge from the ancients because of the Roman bath towns that had been developed around the mountain hot springs in the Pyrenees. The baths were gone, but the hydraulic systems were being modified and used by peasant women for their own purposes, maintaining Roman knowledge of hydraulics but using it to supply water to public laundries and private homes (Mukerji 2008). Languedoc also had many mills, using diversionary canals for millstreams, and rivers that had been partly canalized and equipped with simple locks to bypass rapids. Some regions were also rich in diversionary channels used for irrigation, town water supplies, home water needs, and laundries. The region also had some engineering schools that taught the limited formal knowledge of hydraulics.

One of the faculty members from the school at Castres in the Montagne Noire, Pierre Borel, had even devised a plan to link Languedoc's major rivers with canals to create a navigational system for the whole region. The project only failed when Borel lost his job because he was a Huguenot; landholders would not give up their land to build his infrastructural project. Borel then went to Paris (Blanchard et al. 1985, 181–94).

Riquet had his own lands and a mill in the mountains near Castres, and he recognized the value of Borel's plan for integrating water systems. So the tax farmer, although he was not an engineer, nonetheless proposed building a navigational canal where Borel had imagined one. Riquet knew so little about hydraulics that he probably did not even know how complex a hydraulic system he was proposing or that it was technically impossible with the formal engineering knowledge of the period. He just assumed that Borel knew it was possible, and he believed in the skills of local laborers.

WATER AND THE POLITICS OF IMPERSONAL RULE

The story of the Canal du Midi is an interesting example of the turn to impersonal rule and the recruitment of water into French political administration. The completion of this waterway demonstrated conclusively that impersonal forms of governance could serve as a counter to patrimonial politics, depriving nobles of their control over the region by changing the hydrology.

Riquet was astonished by the political effectiveness of infrastructural change, and bragged to Colbert about how much he was doing for the king. He testified to the power of impersonal rule with this enthusiasm, but it was a political mistake, revealing his naïveté. In patrimonial terms, he was servant of the king, deriving his power to build the canal from the monarch, not giving power to the king. For this indiscretion, Riquet was politically marginalized and was branded by Colbert a dangerous man. But in noticing how the canal project was changing social life around him, he provided evidence of the efficacy of water as a tool of power (Mukerji 2009).

Social historians and sociologists, particularly those influenced by Marxism and world-systems analysis,[2] have underscored the importance of land to power, but most social theories of power do not recognize water as an important political asset.[3] Nonetheless, water supplies have historically been just as necessary to states as land.[4] Moreover, water has numerous interesting properties and uses. Because water is a liquid, it acts quite unlike solids like rock and soil. It flows relentlessly downhill, so it can be used as a source of power for mills. And it floats boats, so it can serve as a medium for transportation. Alternately, it can cause floods, erode structures, destroy cities, and leave behind refuse and mud after a storm because it is an unruly material that defies easy control. So, water management raises distinctive engineering problems and possibilities for social life, and it can be an important source of social change—as it was at the Canal du Midi in seventeenth-century France.

Building the Canal du Midi gained political legitimacy in spite of local hostility to the project for cultural reasons. The restoration of nature through virtuous governance had religious sanction. It was the moral obligation of good Christians and virtuous leaders to restore the landscape to the perfection it had in the Garden of Eden. Making land more peaceful and productive was also deemed possible because Adam was made in God's image with a capacity to know his Creation.[5]

The idea of good stewardship and restoring nature to its Edenic form was powerful in many parts of Europe, and has often been identified as part of Protestant culture. Still in France, *mesnagement* stewardship politics was associated with religious tolerance, and pursued by Catholics and Protestants alike. If theologians disagreed about how to read the Bible, they could agree about how to govern the earth. It was the obligation of all descendants of Adam and Eve to try to undo the damage of the Fall. The English and Dutch used the idea of land stewardship to justify their overseas colonial activity. But the French focused *mesnagement* politics on the restoration of France itself after the ravages of the Wars of Religion. Olivier de Serres (1600) even argued that a king proved he was the true agent of God on earth when he used his God-given intelligence and moral fiber to restore the earth to its true form (Mukerji 2005; 2007a). In light of Serres's writing, Henri IV began to experiment with territorial governance, and Colbert later imitated these policies, considering cutting a navigational canal across Languedoc (Mukerji 2002; Pinsseau 1944).

Because it was a water channel, the Canal du Midi entered politics in an impersonal way. It was an agent of the state that could not be killed, but still had enormous influence over local life. It started to carry the mail, support wine production, link weavers to textile finishers, facilitate trade in leather, and integrate the region more politically. It also interrupted old ways of life, by cutting across roads and fields, destroying orchards and taking away business from mills. It also introduced money and a free labor market into a region that had been dominated by local estates and a peasant economy.

Locals who did not like the changes were faced with a problem because the canal was an agent of the state made of water, not flesh. They could resist the waterway by breaking down the sides of the channel where it was elevated, effectively stopping transport by stranding boats. But water would continue to flow out of any opening in the side of the canal, causing flooding downhill that would damage local towns and crops, while doing nothing to endanger the king or the power of the state. Water in the Canal du Midi became an uncanny sight and evidence of superhuman power, flowing far from any source, hugging the sides of mountains, and meandering through arid land. It stood for the state but far from the person of Louis XIV, illustrating the monarch's capacity to reshape Creation itself to serve as steward of his kingdom.

Exercising power using water eroded patterns of personal power because it worked by changing the context of life, not influencing social actors. It did not rely on social networks and patrimonial relations, so it evaded the techniques of control that regional nobles and clergy had cultivated for centuries. These elites had no obvious weapons with which to fight a conduit of water that spanned kilometers, crossing their land but also moving far beyond it. They had no powers to stop a king who had the right to indemnify lands for his own purposes, had a duty of good stewardship, and could impose new taxes to pay for infrastructural improvements. Under these conditions, water was a powerful weapon, and reshaped the lives of even those who vociferously opposed the will of the king.

IT TAKES WATER TO FLOAT BOATS

To look more closely at how the power of water was captured for the state and the Canal du Midi, I want to focus on the alimentation system that supplied water where the canal crossed the continental divide (Adgé 1992, 202–3). The navigational waterway would only work if it had a hydraulic system to bring water in the requisite quantities to the proper place, providing means for filling locks to raise and lower boats while managing water's unruly tendency to break out of channels or seep into the ground (Mukerji 2009).

A range of projects for a canal across Languedoc had been designed before Riquet proposed his project. The earliest ones were advanced by engineers who came to France from Italy, including Leonardo da Vinci. When François I gained control of Milan in 1515, he visited Lombardy, and was struck by the number and usefulness of the canals there. In 1516, he returned to France with Leonardo, and asked him whether it would be feasible to build a canal to connect the Garonne to the Aude River in Languedoc (Rolt 1973, 13–16; Gazelle 1985).

There was a long east-west valley north of the Pyrenees and south of the *massif central* that could—in principle—be spanned, using the Garonne to the west and the Aude to the east (figure 1.1). It was not as flat as the Lombardian plain, but could conceivably carry a canal. The problem, according to Leonardo, was finding a water supply (Rolt 1973, 16; Gazelle 1985, 147). The water supply was crucial to the project because the canal had to pass from the Mediterranean to the Atlantic watershed, crossing the continental divide or the *pointe de partage des eaux*, as it would be called. At this *pointe de partage*, water drained in two directions toward the two seas, and had to be replaced by water from a higher source. The question was where to get the water and how to deliver it to a canal through central Languedoc.

Riquet understood the problem of the water supply, and apparently liked Borel's solution: routing the canal itself through the mountains. Borel, when he had taught engineering at the university at Castres, had became familiar with a wide plain along the continental divide near Revel, where he thought he could build a canal joining rivers that flowed from this mountainous area in two directions: to the Atlantic Ocean and to the Mediterranean Sea (figure 1.2). This approach was counterintuitive because his canal would not follow the main east-west valley of Languedoc, but

FIGURE 1.1. Valley in Languedoc below Montagne Noire. Photo courtesy Christian Ferrer.

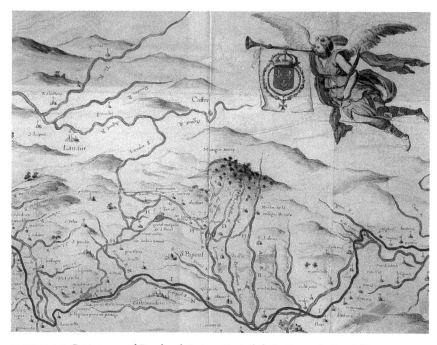

FIGURE 1.2. Region around Revel and Castres. Carte de la Partie . . . du Canal. Photo courtesy le Service Historique de la Défense R21n9A.

instead would veer north into the mountains before Toulouse. But a canal in this area could depend on water supplies from the nearby Montagne Noire.

Riquet never mentioned Borel or spoke of his proposals directly, but he worked on his idea for the canal with the Bishop of Castres, D'Anglure de Bourlemont, and in 1662 wrote Colbert to propose essentially Borel's plan.[6] The proposal interested the minister, but Colbert wanted a commission of local notables and experts to evaluate it. He appointed the Chevalier de Clerville to head the commission, and Clerville, in turn, asked an engineer named Boutheroue, who managed the Canal de Briare near Paris, to work with Riquet to finalize a plan. Boutheroue insisted that the canal had to stay in the main valley of Languedoc and extend to the Garonne River near Toulouse, but he also thought that Riquet could use the route across the Revel plain to bring water from the Montagne Noire to supply the Canal du Midi (Mukerji 2009, 36–59).

This proposal was provisionally approved by the commission as long as Riquet could prove that the water supply was viable. The commission called for the entrepreneur to build a *rigole d'essai*, a smaller, trial version of the water supply system, to show how much water he could deliver from the Montagne Noire to the *pointe de partage* at the *seuil de Naurouze*.[7] When the rigole was completed months later, the commission not only witnessed the arrival of water at Naurouze but was given a *Relation particuliere de la rigolle dessay*, a final report on the project describing how it was built.[8]

THE RIGOLE D'ESSAI

Examining the *Relation particuliere de la rigolle dessay*, one can get a very good picture of the work involved, although it was a bureaucratic document officially reporting to the commissioners what they already knew: Riquet had been able to bring adequate water to the *pointe de partage* at Naurouze. But the *Relation* was not simply verbal testimony, a *procès-verbal* written by men of rank and legal standing to testify to the success of an experiment. It was a *relation*, or a narrative accounting of what was done, why it took so long, what problems were encountered, and how the problems were solved. It was written anonymously by a nameless expert—a man who I think was Riquet's young assistant, Pierre Campmas. Campmas was the son of a local *fontanier*, and the person who had most centrally designed the water supply. Campmas was not a formally educated man, but rather a young recruit into a trade that he was learning from his father. He had neither the social standing nor the experience to give his word authority, but he was literate and knew enough to describe with precision the technical problems and solutions involved in the work.

The author of the *Relation* seems to have been Campmas because he said that people doubted his ideas on account of his being so young, but the rigole d'essai had worked, proving him right. The *Relation particuliere de la rigolle dessay* is also a technical document that describes in detail both the landscape of the Montagne Noire and how it was changed to bring water from high sources over the plain of Revel down to Naurouze. The author focuses on material agents like rocks and water, explaining how they normally interact in the mountain to produce its terrain, and he specifies how their powers were overcome or harnessed to build the rigole d'essai. The narrative of human power and powerlessness in the natural world would make particular sense to a rural *fontanier* whose job was to make recalcitrant water serve human purposes.

In the *Relation*, rocks impede access to places; water does what it wants, including disappearing into sand; soils change from place to place and pose different problems of construction and water-tightness. Riquet and his workers try to change relationships among things. Rocks are moved to let water flow downhill in a new direction. Routes are chosen so water will not flow too fast. The agency of people is used to control the agential properties of water. All the work is done on rocks, sand, and gravel, but the parameters of the work are set by the properties of water. Even the success of the canal builders is measured by the arrival of water from the rigole d'essai at the seuil de Naurouze. Water is preserved as an agential material because it is needed for floating boats and flowing through locks to the seas. Success entails the transfer of that agency to the state. Water is made a tool of impersonal rule.

Riquet began working on the rigole d'essai in May 1665, but progress was slow, so it was only completed in October. There were technical roadblocks and damaging floods that impeded the work. Water and rock kept following their own natures rather than submitting to the will of the king. Delivering water was no easy task because the mountain was wild and its sources were both high and remote. Also, the rigole had to cross the continental divide many times along the way, so keeping the inclines correct—a practice necessary to keep the water flowing—was not easy to assure. Building a watertight conduit in bad land was also no mean feat. So, while water could be diverted from streambeds near sources, what to do with it next was an ongoing problem.

Workers hit new springs in some places, or ran into small underground tunnels that sucked the water away. There were rocks and high scarps that stood between the mountain's high rivers and the main canal that had to be crossed without turning the rigole's waters into a waterfall that would tear the rigole apart. The Montagne Noire and its rivers had accommodated each other over centuries, creating a topography that defined them both. Now Riquet and the commissioners were asking water to follow new paths, and workers were trying to use natural materials to create unnatural effects (figure 1.3). Breaking down the habitual relationships of rocks, soil, and water in the Montagne Noire was an act of hubris. But it was also a means for capturing logistical power for the state, using the power of water to break down local social relations that had kept elites of Languedoc strikingly autonomous from the king.

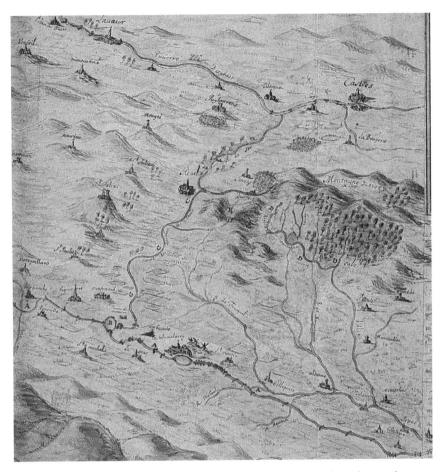

FIGURE 1.3. Proposed captures on the Montagne Noire. Carte du Canal Royal, 1677. Photo courtesy le Service Historique de la Défense R21n30A.

The *Relation particuliere de la rigolle dessay* says that the test system for the water supply started at the Alzau River at a place called Calz. At this remote spot high in the mountains, river water gushed in a steady and strong flow. The location was so remote that it could not even be approached on horseback. Employment records show that Riquet used local men for this work (Adgé 2001), who apparently accessed this part of the mountain on foot. There were massive rocks at Calz that formed a deep ravine filled with sandy soil. There was no obvious way, the verification argued, to cut through the rocks or use the ravine without losing water. So, the crew built a wooden trough, presumably like the ones used for mills, to carry the water over the rocks and down through the ravine.[9]

Riquet, perhaps deciding that the whole rigole should be built this way, wrote to the horrified Colbert about this possibility, asking for permission to acquire masses of timber.[10] But before the minister could say no, the entrepreneur wrote back that he had found another way to proceed. Perhaps the workers started blasting the rocks. The verification document mentions in a later section that they had run low on powder, so perhaps they had been using gunpowder for getting past rocks.[11] What was rare in the mountain was a place where the soil was easy to dig, and a ditch would hold water by itself.

The water in the mountains kept responding to gravity more than the will of the king. If building the rigole was a matter of asserting human agency over the water, governing it through engineering, human agency was losing and the government seemed to have reached its logistical limits. Working with the materials at hand in the mountains, workers kept losing this elusive fluid rather than delivering it to the central valley of Languedoc. Both gravel and sand leaked water, sending it back to the rivers where it wanted to flow. Sand was easier to dig and could produce a nicely shaped ditch that could be reinforced with pilings, but often the rigole walls collapsed like sugar cubes into the currents when water started to flow. Sand often seemed to flow like a liquid along with the water it was supposed to contain.

Rocks made the terrain hard to traverse to get access to sources and impeded the construction of the rigole. Where there were large boulders or scarps, there was no hope of removing or moving them. Routes could be blasted through them using gunpowder, but this was hard, slow, and expensive work. The granite of the Montagne Noire paid the dividend, however, of providing strong, watertight material for the conduits. Sometimes workers used natural riverbeds as part of the rigole, adding more water from mountain sources and taking it out later where the conditions were less taxing. Many techniques were tried because the problems were varied and the inclines had to be precise no matter what type of terrain needed to be crossed. The rigole d'essai was a struggle with the solid materials of the mountains, but the parameters for its design were set by the demands of water itself.

Not all the technical problems were solved to create the rigole d'essai. The structure was provisional and remained leaky. The permanent rigoles made later were more watertight, lined in many places with a layer of pounded clay. Still, wooden pilings and planks, blasted rocks, and high berms shored up the experimental structure well enough to bring water in large quantities from the Montagne Noire across the plain of Revel and to the seuil de Naurouze.

In only one part of the *Relation* are workers criticized for the poor quality of their efforts. If the mountain and its materials created the problems in capturing water from high sources, faulty surveyors created the problems of routing on the plain of Revel. On this plain, the channel had to cross from the Atlantic to the Pacific watershed, but following the prescribed route, the water did not flow where the rigole had been dug. If most of the story in this report described human agents triumphing against the unruly forces of nature, at this point, where

FIGURE 1.4. Area of the rigole d'essai. Andreossy map, 1665 Rigole de la plaine. Photo courtesy le Service Historique de la Défense R21n5A.

human intelligence was most necessary, it failed. But common sense prevailed, as the rigole was routed along millstreams that also crossed this divide.

Once the rigole left the plain of Revel, it passed down a long valley with good soil that was easy to dig, held its shape, and did not leak significant water. Attention to elevation remained crucial, since the water had to move by gravity feed alone toward the canal at the continental divide. The slopes were gentle, but the valley had hillocks to navigate around. In some places, the rigole was elevated with stone, wooden, and dirt berms to maintain its incline and remain above the level of the sieul de Naurouze (figure 1.4).

Along the way, the rigole picked up more water from local sources, helping restore some of what was lost on the mountainside. Naurouze itself had water sources that added to what was collected in the reservoir for the canal. So, by the time the rigole d'essai reached Naurouze, it brought a massive flow of water to this spot, a man-made river of sorts, that had circled down from the mountains, flowed into rivers and out, and skirted across the plains to provide alimentation for the Canal du Midi.

The test water supply system provided a calculable input of water into the Canal du Midi that was judged nearly adequate in itself for navigation in the dry summers of Languedoc. About a third of the water was thought to be lost but mostly recoverable in a permanent rigole. This proved to be an exaggeration, and the canal sometimes had to be closed in summer because of a lack of water. Still, the rigole d'essai was a massive success, and this assessment allowed Riquet to receive a contract for construction of the Canal du Midi.

The *Relation* ends with some congratulatory and celebratory remarks about Riquet's success. The report emphasizes that no one had really had a full sense of the complexity of the project of building the rigole, but that the concept remained correct and was shown to be viable. The author asserts that the verification vindicated him by demonstrating that his proposal was an honest one based on true knowledge. It shows him to be a person of honor, not just ability. He argues that he would not have proposed a project that was not feasible.

The final comments are an assertion of agency that associates the author's capacities to realize the rigole d'essai with the power of the human mind to exercise dominion over nature. Campmas, if he is the author, speaks in a language familiar to stewardship politics about the nature of logistical power. Forcing water from the mountain and taking it to Naurouze was an act of human dominion, and using it to build a peaceful waterway was an act of restoration, using the human intelligence given by God to Adam. Exercising logistical power, in this context, was not just a way to control people through the control of things, but a moral act of political efficacy. Descendants of Adam and Eve were supposed to tame wild nature and make it more serviceable, using Creation wisely.

The author defends his moral standing and personal honor by showing how Riquet's workforce made the water supply successful. Riquet and his workers could only exercise logistical power for the king by recognizing the properties of natural things like water and rock, using materials in the mountains to control the agential properties of water and make them serve a navigational canal to join the two seas. This was an act of stewardship.

The report was necessary to document to the commission how the system was built, but it was not the measure of its success. Success was demonstrated by material means with the flow of water at Naurouze. Everyone who saw the water arrive at this pass over the continental divide understood that Riquet's plan was viable. The blow-by-blow account of the hard work and technical difficulties involved only demonstrated the heroic dedication of Riquet and his men to make the water of the Montagne Noire serve Louis XIV.

CONCLUSION

Building an effective hydraulic infrastructure was a means of asserting and legitimating a new kind of power—logistical power. France's weak state, which had for so long been unable to get local people to submit to the king, was now altering life in Languedoc in consequential and permanent ways. The water system demonstrated the impact of territorial governance, using natural forces to exercise the will of the king. The exact properties that made the water in the Montagne Noire so difficult to contain and deliver to the Canal du Midi also made the Canal du Midi a powerful force in the region. Once water started flowing through the canal, it was hard to stop.

Building the Canal du Midi was also a practice of impersonal governance that changed local life, shifting power away from local noble officials and landholders to agents of the state. Mills on rivers lost business to new ones along the canal. The roads managed by nobles were replaced for transport by the king's waterway. The mail came by boat on the canal; women washed laundry in it; and merchants traded in textiles, leathers, and wine. All this entangled local elites with the state administration that they had ignored before. Breaking apart rocks in the mountains and lining sandy conduits with clay put into the hands of the state a new capacity for shaping social life that was startlingly novel and powerful. Now the northern monarchy was not a distant distraction, but an unrelenting presence in Languedoc.

The Canal du Midi instantiated a new kind of power of the state: something superhuman, uncanny, and daunting—a technique of impersonal rule (Mukerji 2009, 154–75). It was a work of logistical power, managing social relations by material means, and it was an impersonal instantiation of what a state could do. People in Languedoc may have understood that building the canal was the will of the king, but they were also witnesses to the physical labor and exercise of natural knowledge that made the canal work. And they were the ones whose lives began to change as nobles became landlords, renting land along the canal that renters planted with vineyards to make wine for export, and as new towns developed to take advantage of increased trade. The inhabitants of these towns and the cities along the canal were the ones who did their laundry in the canal's waters and found jobs in the new businesses, breaking down the peasant economy to replace it with a money system. Patterns of everyday life changed (Carroll 2006; Joyce 2003; Parker 1983; Scott 1998) without anyone telling people what to do. Logistical power worked silently outside social pressures and, in this sense, seemed apolitical even as it initiated social change.

The story of the reign of Louis XIV, particularly the tale of the growth of state absolutism, is usually told in terms of the king's life at Versailles (Apostolidès 1981; Neraudau 1986; Elias 1998; Mukerji 1997). But Louis XIV gained greater power over Languedoc through an exercise of stewardship of massive proportions that seemed to support the legitimacy of the northern monarchy. He seemed to be capable of fulfilling, too, France's destiny in restoring the glory of Gaul. The French were clearly able to engage in the kind of massive engineering work that had made Rome so powerful. This was not the only engineering to be touted in their period. There were the massive fortresses of Vauban and the new port cities along the Atlantic coast. France was changing, and so French life had to shift. Engineered by the state itself, life became French to fit the territory of the king. In this sense, the empowerment of the French state in the late seventeenth century is a story of impersonal rule, not personal rule. The state made Louis XIV seem absolutely powerful because it exercised powers that were not fully understood. It exercised the power of forestry, fortress engineering, archiving, book collecting, and bookkeeping. And in all these cases, politics was depersonalized as the agential properties of things were put to political use. Trees, stone, water, rags, and ink were the secret powers of administration that Colbert elaborated, and these were the impersonal tools he used to make the will of the king seem absolute.[12]

ACKNOWLEDGMENTS

Thank you to CASBS and UCHRI for funding in support of this research.

REFERENCES

Adgé, Michel. 1992. "L'art de l'hydraulique." In *Canal royal de Languedoc: Le partage des eaux*. Conseil d'architecture, d'urbanisme et de l'environnement de la Haute-Garonne; Caue: Loubatières, 202–3.

———. 2001. "Premier Etats du Barrage de Saint-Ferréol." *Les Cahiers d'Histoire de Revel* 7.

Adkins, Lesley, and Roy Adkins. 1994. *Handbook to Life in Ancient Rome*. New York: Oxford University Press.

Anderson, Perry. 1974. *Lineages of the Absolutist State*. London: NLB.

Apostolidès, Jean-Marie. 1981. *Le roi-machine: Spectacle et politique au temps de Louis XIV*. Paris: Éditions de Minuit.

Appuhn, Karl. 2009. *A Forest on the Sea: Environmental Expertise in Renaissance Venice*. Baltimore: Johns Hopkins University Press.

Bartoli, Michel. 2011. "Louis de Froidour (1626–1685), Notre Héritage Forestier." *Les Dossiers Forrestiers* 23.

Beik, William. 1985. *Absolutism and Society in Seventeenth-Century France: State Power and Provincial Aristocracy in Languedoc*. Cambridge: Cambridge University Press.

Bergasse, Jean-Denis, ed. 1982–86. *Le Canal du Midi*. 4 vols. Cessenon: J.-D. Bergasse.

Blair, Ann. 2020. *Too Much to Know: Managing Scholarly Information before the Modern Age*. Hew Haven: Yale University Press.

Blanchard, Anne. 1979. *Les ingénieurs du roy de Louis XIV à Louis XVI*. Montpellier: Université Paul-Valéry.

Blanchard, Anne, Michel Adgé, and Jean-Denis Bergasse. 1985. "Les ingénieurs du roy." In *Le Canal du Midi*, edited by Jean-Denis Bergasse, 4:181–94. Cessenon: J.-D. Bergasse.

Brewer, John. 1989. *The Sinews of Power: War, Money and the English State, 1688–1783*. New York: Knopf.

Carroll, Patrick. 2006. *Science, Culture and Modern State Formation*. Berkeley: University of California Press.

Cole, Charles Woolsey. 1964. *Colbert and a Century of French Mercantilism*. 2 vols. London: Cass.

Cronon, William. 1992. *Nature's Metropolis: Chicago and the Great West*. New York: W.W. Norton. de Froidour, Louis. 1899. *Les Pyrenées Centrales au XVIIe siècle: Lettres écrites . . . à M. de Héricourt . . . et à M. de Medon*. Auch: G. Foix.

Dent, Julian. 1983. *Crisis in Finance: Crown Financiers and Society in Seventeenth-Century France*. New York: St. Martin's Press.

Drayton, Richard. 2000. *Nature's Government: Science, Imperial Britain, and the "Improvement" of the World*. New Haven: Yale University Press.

Elias, Norbert. 1998. *The Court Society*. New York: Pantheon.

Froidour, Louis de. 1899. *Les Pyrenées centrales au XVIIe siècle; lettres écrites ... à M. de Héricourt ... et à M. de Medon*. Auch,: Impr. et lithographie G. Foix.

Gazelle, François. 1985. "Riquet et les eaux de la Montagne Noire: L'idée géniale de l'alimentation du canal." In *Le Canal du Midi*, edited by Jean-Denis Bergasse, 4:143–70. Cessenon: J.-D. Bergasse.

Gébara, Chérine, and Jean-Marie Michel. 2002. "L'aqueduc romain de Fréjus: Sa description, son histoire et son environnement." *Revue Archéologique de Narbonnaise*, supplement 33. Montpellier: Éditions de l'Association de la Revue Achéologique de Narbonnaise.

Holt, Mack P. 2001. *The French Wars of Religion, 1562–1629*. Cambridge: Cambridge University Press.

Joyce, Patrick. 2003. *The Rule of Freedom: Liberalism and the City in Britain*. London: Verso.

Kettering, Sharon. 1986. *Patrons, Brokers and Clients in Seventeenth-Century France*. New York: Oxford University Press.

Lansing, John. 1991. *Priests and Programmers: Technologies of Power in the Landscape of Bali*. Princeton: Princeton University Press.

Long, Pamela O. 2001. *Openness, Secrecy, Authorship: Technical Arts and the Culture of Knowledge from Antiquity to the Renaissance*. Baltimore: Johns Hopkins University Press.

Lynn, John. A. 1997. *Giant of the Grand Siècle: The French Army, 1610–1715*. Cambridge: Cambridge University Press.

Maistre, André. 1998. *Le Canal des Deux-Mers: Canal Royal du Languedoc, 1666–1810*. Toulouse: E. Privat.

Mann, Michael. 1986. *Sources of Social Power*. New York: Cambridge University Press.

Miller, Peter N. 2000. *Peiresc's Europe: Learning and Virtue in the Seventeenth Century*. New Haven: Yale University Press.

Minard, Philippe. 1998. *La fortune du colbertisme: État et industrie dans la France des Lumières*. Paris: Fayard.

Mousnier, Roland. 1979. *The Institutions of France under the Absolute Monarchy, 1598–1789: Society and the State*. Translated by Brian Pearce. Chicago: University of Chicago Press.

Mukerji, Chandra. 1997. *Territorial Ambitions and the Gardens of Versailles*. Cambridge: Cambridge University Press.

———. 2002. "Material Practices of Domination: Christian Humanism, the Built Environment and Techniques of Western Power." *Theory and Society* 31: 1–34.

———. 2005. "Dominion, Demonstration and Domination: Religious Doctrine, Territorial Politics, and French Plant Collection." In *Colonial Botany: Science, Commerce, and Politics in the Early Modern World*, edited by Londa Schiebinger and Claudia Swan, 19–33. Philadelphia: University of Pennsylvania Press.

———. 2007a. "Demonstration and Verification in Engineering." In *The Mindful Hand: Inquiry and Invention from the Late Renaissance to Industrialization*, edited by Lissa Roberts, Simon Schaffer, and Peter Dear, 169–88. Chicago: University of Chicago Press.

———. 2007b. "The Great Forest Survey of 1669–1671." *Social Studies of Science* 37 (2): 227–53.

———. 2008. "Women Engineers and the Culture of the Pyrenees." In *Making Knowledge in Early Modern Europe: Practices, Objects and Texts, 1400–1800*, edited by Pamela Smith and Benjamin Schmidt, 19–44. Chicago: University of Chicago Press.

———. 2009. *Impossible Engineering: Technology and Territoriality on the Canal du Midi*. Princeton: Princeton University Press.

Murat, Inès. 1984. *Colbert*. Translated by Robert Francis Cook and Jeannie Van Asselt. Charlottesville: University Press of Virginia.

Neraudau, Jean-Pierre. 1986. *L'Olympe du Roi Soleil*. Paris: Société des Belles-Lettres.

Parker, David. 1983. *The Making of French Absolutism*. London: Edward Arnold.

Pinsseau, Pierre. 1944. *Le Canal Henri IV ou Canal de Briare (1604–1943)*. Orléans: R. Houzé.

Rolt, L.T.C. 1973. *From Sea to Sea: The Canal du Midi*. London: Penguin.

Scott, James. 1985. *Weapons of the Weak. Everyday Forms of Peasant Resistance*. New Haven: Yale University Press.

———. 1998. *Seeing like a State: How Certain Schemes to Improve the Human Condition Have Failed*. New Haven: Yale University Press.

Scott, Joseph Frederick. 2006. *The Scientific Work of René Descartes, 1596–1650*. Ann Arbor: University of Michigan Press.

Serres, Olivier de. 1600. *Le Théâtre d'Agriculture et Mesnage des Champs*. Paris: I. Métayer.

Soll, Jacob. 2009. *The Information Master: Jean-Baptiste Colbert's Secret State Intelligence System*. Ann Arbor: University of Michigan Press.

Tilly, Charles. 1975. *The Formation of National States in Western Europe*. Princeton: Princeton University Press.

Wallerstein, Immanuel. 1974. *The Modern World System*. New York: Academic Press.

NOTES

The documents cited here (ACM) are from the Archives du Canal du Midi operated by the Voies Navigables de France in Toulouse, and managed by the archivist, Samuel Vannier.

1. Most writers on Colbert emphasize his economic policies to the exclusion of his other experiments in material governance. See, for example, the classic: Cole (1964). See also Murat (1984); Minard (1998); and Dent (1983).

2. See, for example, Carroll (2006); Joyce (2003); Mann (1986); Parker (1983); Scott (1998); Tilly (1975); and Wallerstein (1974).

3. Michael Mann (1986) even speaks of logistics.

4. See Lansing (1991) and Cronon (1992). See also the interconnectedness of water-based power and land control in Appuhn (2009).

5. See, for example, Drayton (2000).

6. Mukerji (2009, 43–48); Rolt (1973, 24–26). Pierre Borel went on to Paris when he left Languedoc and became quite well known. He entered the Académie Royale des Sciences as a chemist and wrote on Cartesian science. See Scott (2006, 84).

7. ACM 2–14; Mukerji (2009, 56–59); letter from Riquet to Colbert, August 18, 1665, ACM 20–21.

8. ACM 2–14.

9. ACM 2–14, pp. 1–2.

10. Letter from Riquet to Colbert, September 3, 1665, ACM 20–19.

11. ACM 2–14, p. 11.

12. Blanchard (1979); Bartoli (2011). Compare to Appuhn (2009).

2

Winnipeg's Aspirational Port and the Future of Arctic Shipping (The Geo-Cultural Version)

Stephanie C. Kane

Ethnographers usually study ice/water and culture in contemporary and historical timespaces. They tend to start somewhere between present tense and time immemorial of Indigenous peoples. This corresponds to a tiny sliver of timespace in geological terms. But if ethnographers and humanities scholars more generally want to understand humans as geological actors with powers and ethical responsibilities to influence planetary ice/water futures, then we must invent ways to stretch culture into geo-culture.[1] Most human actions causing break-up and melting of Arctic sea ice, among other effects and drivers of climate change, have occurred within the last one hundred years or so.[2] However, I believe that ethnographers might think about humans as we think about glaciers—place-based and place-transiting collective, embodied agents who act in processes that extend back into the last ice age.

Representations of water (and water bodies) generally tend to be dynamic but inert, or at least always imagined as bound by Nature's laws to act accordingly.[3] Ethnographic waters are universally required and manipulated, ritualized, rerouted, and polluted by *people* whose actions focus whatever dreams and dramas are at stake.[4] But I wonder, how might the drama of a melting Arctic Ocean be plotted as one in which agencies are shared more equally between humans and water? Or, better still, a drama in which both humans and ice/water have agential powers that form and act materially and culturally on the stages of Earth's amphibious crust?[5]

In this chapter, I experiment with a geo-cultural approach that focuses on collective actors in global shipping. Global shippers have contributed unintentionally and indirectly, but nevertheless certainly, to Arctic sea ice loss. As they now approach widening passages through retreating sea ice as opportunities to facilitate more efficient global trade, in the process, their fossil-fuel-burning, ice-breaking ships may accelerate sea ice loss. The city of Winnipeg in Manitoba, Canada, with its oddly placed aspirational port and glacial destiny, is the focal point for this geo-cultural experiment.

Diverse human collectives often come together to reconfigure and integrate various water bodies, altering their direction of flows and forging new intercon-nections among them. The engineered aquatic thing that results demands con-tinuing attention from people who would attempt to borrow its essential power. In *Impossible Engineering*, her history of a seventeenth-century French interoceanic canal, the Canal du Midi, Chandra Mukerji (2009) opens up an infrastructural path towards narrating water's agency. Mukerji writes (204):

> The waterway became both an unquestionable fact of local life like a mountain or a wild river, laying down the conditions of possibility for all those living near it, and the opposite: a product of collective action and an ongoing raison d'être for social coordination.

A fact of life and a product of collective coordination: these are existential fea-tures of a social relationship in which water lends its "logistical power" to humans and in so doing organizes (and sometimes thwarts) their capabilities and mobili-ties. Mukerji (2009, 214) distinguishes logistical power conferring dominion over things (such as water) from strategic power conferring dominion over people. Logistics, "or the mobilization of natural forces for collective purpose," brings "different types of power to social life." The concept of logistical power frees this author from the claustrophobic struggle between interest groups to explore pos-sibilities for a political ecology of water that takes water's agency, or waters' agen-cies, seriously.

What I take away from Mukerji's history is the idea that water's power, as people attempt to move it this way and that through varied configurations of rock and soil, has to work through an appreciation of its changeable, interactive nature. Human intention emerges in place and in process. People have to feel their way along as they navigate the conditions of possibility that water bodies allow. To achieve this sense of the entwined agencies of humans and water in geological timespaces, I find it useful to broaden our view of human agency, to actively acknowledge and explore unintended spheres of even our most intentional actions.[6]

In the spheres of unintended effects, and affects, I propose that agencies of water and humans interact more equitably. Once one moves human intention into the background, shocking stories of climate change, such as melting of the Arctic Ocean as existential hazard and capitalist boon, can be told otherwise, and per-haps better questions can be asked.

Human collectivities, some more than others, have caused climate changes that have dramatically altered human-water relationships. A significant *cultural* fact attends this scientific fact: humans have *unintentionally* caused climate change.

Scientists have identified intentional actions that drive climate change—fossil fuel use the most prominent cause and Arctic sea ice loss a prominent effect. Humanities scholars, however, may well be better positioned to explore significant unintentional drivers of anthropogenic climate change. They can do so by ceding the focus on intentional human agency in every narrative, sharing agency with water.

Integrating the scientific fact of climate change into culture is difficult. Toward this end, I do believe that (re)writing human-water interactions within the sphere of unintentional agencies could dampen anthropocentric bias and lead to circulation of other kinds of climate change narratives. I do this analytically here by carrying port city ethnography into geoscience research that illuminates how ice/water moves in and across Earth's crust. When I poach geoscience narratives that precede human encounter, yet a second door out of the human-struggle-interest box opens. Through it, I try to find my way into a humbler, yet scientifically allied sense of *the real* that stretches the future back before time immemorial.

This chapter takes up this book's aquatic humanities theme by tackling a sphere of unintentional agencies relating to maritime engineering, ice/water geology, and the city. The chapter has three main sections and a conclusion. The first two sections shift analysis of human-water relationships in new directions using two experimental maneuvers. The third brings it all together in Winnipeg. First, I articulate a double human-water maneuver that shifts conventional water discourse. On the human side of this relationship, I shift away from particulars of our intraspecies collaborations and confrontations and toward meso-scale collective actions that create port cities and monumental water engineering projects.[7] And on the water side of the relationship, I shift away from abstract flows and cycles and toward meso- and macro-scale *water bodies*, such as river basins, lakes, and oceans. I flesh out this double human-water maneuver by examining Winnipeg's aspirational port through a historical lens of port cities, interoceanic canals, and other key technological revolutions in the global transport industry.

Second, I attempt to recast the meaning of the term *anthropogenic* so as to downplay the significance of human intention in the understanding of climate change. To accomplish this, I mine the published work of geoscientists, who thus become some of my most important key informants. Their work allows me to expand timespaces of human encounter into the geological scale, expanding narrative from the sliver of human presence into the Quaternary Period. In classical geology, the Quaternary begins in the ice age of the Pleistocene Epoch, continues into the current Interglacial, the Holocene Epoch. The Holocene climate melts ice age glaciers to create the major North American water bodies that govern contemporary human activity in and beyond the continent. This second maneuver—qualifying our sense of the anthropogenic by highlighting the power of ice/water—synchronizes our sense of the timespaces of human unintentional actions and effects with the timespaces of water's unintentional actions and effects leading up to the Anthropocene and its various near-end-Holocene variants. In other words, I mine geoscience for visual and narrative data that animate ice/water dramas before and after humans emerged and evolved. By resignifying the role of the unintentional in the anthropogenic, I cede agency to ice/water.

In the third section, narrations of port city history and ice age geoscience converge in the peculiar matter of Winnipeg's aspirational port. CentrePort is peculiar in two senses: for one, it is nowhere near a seacoast (hence its geographically appropriate name). And too, as mostly assigned space and infrastructure waiting to catch up to an idea, CentrePort remains culturally and geographically marginal to its city.[8] Conceptually, however, the anticipation upon which it builds enters into geo-cultural analysis in compelling ways.

The inspiration for this particular piece of my ethnographic puzzle comes from a curious cartographic coincidence. In analysis of materials collected in my study, visual data from CentrePort public relations crossed paths with that of geoscience. The odd intersection winds up crystallizing my argument that water's agency can best be realized at the scale of the melting Arctic Ocean (which is also the planetary scale of climate change and global shipping). That the coincidence appeared at all, I suppose, is evidence in favor of a free-wheeling impulse to roam across interdisciplinary spaces among humanities scholars. Humanities-style roaming opens up new pathways for systematic inquiry, lending imaginative power to scientific knowledge. This is not about illustrating received scientific knowledge. It is about thinking science otherwise.

In analysis to follow, I treat the archive of technical knowledge in geoscience as cultural knowledge because, after all, geoscience informs engineering and engineers lead collective human efforts to rearrange water bodies and human bodies on city, region, oceanic, and planetary scales. I call this realm of inquiry *geo-culture*.[9] Through the geo-cultural, ethnographers can join other humanities thinkers as they "lead diverse scholars and publics into uncertain environmental futures" (De Wolff and Faletti, this volume).

PORT CITIES AND OCEANIC CONNECTIONS (GEO-POLITICS)

As nodes of global trade, port cities in lower and middle latitudes have played an outsized role in organizing the fossil-fueled industrial processes that bring us to the brink of Holocene's end and have put at risk the stability of this geological epoch. Set most often in ecologically rich habitats where fresh water rivers flow into their seas, port cities also host the diverse cultural conjunctures that inspire new forms of human inventiveness, resistance, and resilience (Kane 2012). For better or worse, port cities and routes between constitute the infrastructural and architectural armature upon which, and within which, globalization processes lay their intricate and far-flung foundations, generating a suite of pervasive ideas, objects, and effects mobilized by trade.

The historic opening of shipping routes across the Arctic follows these earlier dramatic breakthroughs in maritime engineering. These breakthroughs altered relationships between major water bodies for the efficiencies of global trade, enhancing the logistical powers of nation-states and multinational corporations. Among the monumental engineering works that heralded modernity are the canals of the nineteenth and early twentieth centuries: the opening of the Erie Canal in 1825 cut through the earth to bring ships between New York and the Great Lakes, opening up the interior of North America to trade; the opening of the Suez Canal in 1869 eliminated the need to go around Africa when, for example, sailing from India to England; and the Panama Canal in 1914 shortened routes between Europe and Asia by connecting the Atlantic and Pacific Oceans.

Reorganizing Oceanic Links: The View from Winnipeg

Engineering feats always produce winners and losers. For the city, the Panama Canal was a disaster, abruptly bringing a thirty-three-year economic boom to a crashing halt. Winnipeg had benefited from the arrival of railroad tracks linking its midcontinental location with eastern Canada in 1881, which enabled the city to become a supply depot for the rest of western Canada. With a diverse population of immigrants arriving between 1881 and 1914, Winnipeg's population grew from 9,000 to 150,000, a growth rate outpacing Chicago. But once the Panama Canal opened the cheaper sea route, it devastated Winnipeg's railway-transport business. The city became so terribly poor that it missed the urban redevelopment phase suffered by many other cities, thus leaving beautiful turn-of-the-century buildings in the Exchange District standing, to become renovated heritage architecture housing Winnipeg's thriving arts and culture scene today (figure 2.1) (Kives and Scott 2013, 38–40). And so it goes when monumental canals redesign global shipping routes by linking major water bodies in new ways: some port cities win, some lose, and some of the losses turn into unexpected boons later on. With such changes in the connections between major water bodies and accompanying forms and routes of human mobilities come pressures to change previously established hierarchies of maritime development in port cities (Kane 2012, 63–65).

Transporting this trend of infrastructure-driven change into the late twentieth century (1970s–90s), the global-scale transition to container shipping irrevocably altered shipping routes once again (Levinson 2006). This revolution grew, the story goes, from one trucker's insight that designing the hold of ships to carry stacks of containers designed to fit the size of truck beds would dramatically simplify port operations. Ports that could secure great investment no longer needed longshoremen. Redesigned docks facilitate loading containers gripped by computer-mediated cranes directly back and forth from truck beds and rail cars to ships. Thus, multimodal or intermodal freight transport was born: the technological innovation that makes Winnipeg's aspirational port possible.

Every jump in containership size and accompanying technological innovation reorganizes the dynamic *flows* of port city waters, ships, people, commodities, and

FIGURE 2.1. Contemporary renovation of heritage architecture in Winnipeg's Exchange District was made possible by the sudden diversion of global shipping through the new Panama Canal about one hundred years before. Photo by author.

information. Reorganization, however, always runs up against port city *fixities* (Desfor et al. 2011; Kane 2012, 118–22). From winter ice that blocks winter passage into harbors, to nineteenth-century stone and concrete docks too small for containerships, to frictions of race- and class-based inequalities that continue to order acts of dispossession and financial reward (see Tsing 2005), fixities shape dynamics of flows. So, if Winnipeg is a bellwether of future industrial fixities meeting climate change flows, port cities that link overseas producers to consumers may not even need to have their own waterfronts, whereas many coastal port cities that now enjoy thriving waterfronts may be under water.

According to the International Maritime Organization, international shipping transports more than 80 percent of global trade.[10] How they move across the oceans matters: unlike the Panama and Suez canals, whose innovation was that they connected oceans and *shortened* shipping distances, containerization by contrast led to *longer* shipping distances. And while containerization was a revolution of efficiency in the transport industry, especially as it became increasingly supported by digital information technologies, it also enabled just-in-time production. Products could be assembled anywhere and everywhere. Together, the container revolution and structural manipulation of labor costs and protections triggered a global displacement of manufacturing. Manufacturing moved from Western industrialized countries, where factories closed and some major ports declined, and moved to Asia, where factories opened and ports expanded. Thus, containerization fundamentally restructured the nation-based political economies of the world (Mah 2014, 1–3; Cowen 2014; Mezzadra and Neilson 2019).

Exceedingly long distances between Asia and Europe or the United States and the accompanying financial and environmental costs of petroleum-based containership transport drive multinational corporate interest in Arctic Ocean routes. Like the canals of the nineteenth and early twentieth centuries, a melted Arctic offers shorter sailing distances. But unless ships sailing Arctic waters use clean energy, the reduced fossil fuel benefits of shorter distance will be offset by increases in atmospheric carbon. In and beyond the Arctic, atmospheric carbon intensifies warming temperatures. In addition to entering global air currents, wherever ships pass, their carbon pollution falls onto and darkens local ice, thereby lowering its sun-reflecting power (albedo).[11]

In any case, like canal construction, new passages through melting sea ice alter interoceanic flows in ways that appear to enhance efficiencies of global trade. But appearances can be deceiving. The logic of fuel efficiency ignores disruptions of hydrological patterns that may accompany melting sea ice. A melting Arctic can induce instability of cold-warm oceanic and atmospheric currents, which may shift Holocene patterns, increase the intensity and unpredictability of storms and storm surges, and hasten loss of coastal terrain and islands upon which port cities stand. Negative impacts on global trade and other forms of geopolitical chaos may attend these and other hydrological impacts of climate change.

My aim here is not to derail optimism among Winnipeg's port boosters betting on Arctic ice loss. Rather, I wish to analyze the Arctic shipping schema as a geo-cultural phenomenon driven by human intentional action directed at oceanic shipping routes and nodes. The schema fits the internationally accepted concept of oceanic space as open, empty, terra nullius. The concept is enshrined in the 1982 United Nations Convention of the Law of the Sea (UNCLOS): open waters are free for all to cross. As Steinberg (2001) has shown, however, to many of the world's peoples, even oceans have histories; for many, oceans are not empty spaces, not merely not-land.

Since nigh the beginning of the Holocene, the Arctic has been home to Indigenous peoples. How will ships plying the waters of a melting Arctic change their lives? As I learned from colleagues on the Ice Law Project, international law is Western law that occasionally entertains some exceptions for rights of Indigenous peoples and has one exception for sea ice.[12] In UNCLOS, Article 234 refers to ice weaving in and under the Canadian Arctic Archipelago. But UNCLOS doesn't account for Indigenous peoples like the Inuit, whose sovereign territory is the sea. For Inuit, whether ice or water, the Arctic Ocean that flows among the islands is part of one navigable habitat for living.[13]

In its hardest, most durable forms, horizontal extensions of Arctic sea ice create solid terrain upon which Indigenous people dwell and navigate the cold ecologies of their subsistence and their cultures. Indigenous knowledge of sea ice reveals it to be an essential substance that morphs, moves, and seasonally recurs and subsides. For Inuit, ice requires nuanced interpretation to understand and survive (Krupnik et al. 2010). Indigenous knowledge shows sea ice to be fundamentally different from land in a way that confounds any simplistic relationship that Western legal instruments, like UNCLOS, posit between land and water (see Aporta 2009; Aporta, Kane, and Chircop 2018; and the Ice Law Project).[14] Yet, still today, unlike Indigenous peoples who, for example, have the right to protect forests, Inuit still have no right to *protect ice* as an environmental good or cultural resource in international law. As Phil Steinberg put it, the problem is that "everyone has a right to break ice."[15]

The fact is, except for summers, the Arctic Archipelago is still icebound, and even in summer there's enough ice around to make shipping hazardous in various ways (Arctic Council 2009; Chircop 2016, 41; Steinberg et al. 2022). If shippers from lower latitudes make no ethical or ecological distinction between functional qualities of sea ice and seawater, they may have no compunction in using icebreaking ships to force open canals. Nineteenth-century canal-building logic of transport efficiency repeats itself. Now targeting a uniquely fragile and complicated Arctic terrain, global shippers wield maritime engineering with regulations governed by a practically ice-blind legal regime.

The first suite of modern megaprojects, the Erie, Suez, and Panama Canals, were, as Ashley Carse (2014, 84–85) writes, "arguably even more symbolically charged than other transportation infrastructure like roads and railroads" precisely because the facilitation of movement "entailed the transformation of the earth itself." The symbolic charge associated with the historic human-built openings dug out to link large-scale water bodies "scrambles" uneasily with lower latitude aspirations that rely on exploiting ice-melted openings (Dodds and Nuttall 2016). In the historical trope, human engineering triumphs over nature. In the current climate-change-aware trope, however, human engineering unleashes destructive forces, some aspects of which become framed as new opportunities. This chapter tells a different story. Humans may take advantage of climate change, but they did not intend, could not have intended, to make climate change happen. In geological timespaces, only narratives that travel into spheres of unintentional agencies can make sense out of the off-kilter nature of the current existential crisis—what Amitav Ghosh (2016) calls *The Great Derangement*.

GLACIAL ORIGINS OF HOLOCENE RIVER BASINS
AND OCEAN CURRENTS (UNINTENTIONAL
HISTORIES IN GEOSCIENCE TIMESPACE)

Geologists are quite comfortable traveling in spheres of unintended agencies. Their technical knowledge can be mined as narrative resources to help tell stories of planetary pasts, presents, and futures. They study the strata of Earth's crust, seeking, recognizing, debating, and naming turning points in time's linear progress (Schneiderman 2015). Dramas of the most recent geological period, the Quaternary, include only the last 1.6 million years before present (b.p.) of glacial comings and goings. Within the Quaternary period, the first epoch is the Pleistocene, which starts when increasing cold envelops the planet and causes the last in a series of ice ages.

In the Pleistocene Arctic, thousands upon thousands of years of snows pile upon snows; the weight of the new falls on the old, making ice; the ice grows great lobes that push out and down across the Northern Hemisphere as glaciers. Extinctions of many large mammals and birds take place as the glacial lobes mass into the Laurentian Ice Sheet and move down into and over North America. As they move, rocks and gravel sediment stick to the glacier bottoms and carve inscriptions into the revised surfaces they push across, registering the mass movements of glacial power that will become clues for geologists.

A warming climate marks Pleistocene's end. Glaciers retreat back to the Arctic from whence they came, and rivers rush out of their bottoms as they melt. Rivers follow the glaciers' retreating path, their meltwaters gathering and swirling with a northward impulse. But they cannot flow through to join ocean currents because the glaciers that give them life block their exits. So, they converge in the center of the continent to form Lake Agassiz. Eventually, ice dams give way and Lake Agassiz bursts out of its earthen boundaries. The stages of glacial demise and their more nuanced fluxes leave patterns in once-submerged prairieland mud (which Winnipeggers call "gumbo"). About 11,700 b.p., the Pleistocene ends at the boundary of this warm period between glaciations, giving way to a new Interglacial era conducive to the development of humankind. Interglacial events inspire geologists to name a second epoch in the Quaternary something wholly new, the Holocene.

In North America, the Holocene marks the slip from life without humans to life with humans, when time immemorial and then history-as-we-know-it begins to unfold in geological timespace. In the Holocene, Indigenous peoples disperse across the continent. Many would gather seasonally at the confluence of the Red River and the Assiniboine River, in the center of a flood bowl once mostly covered by Lake Agassiz. Thousands of years would pass before Anglo-European settlers would establish the province of Manitoba as a political entity with Winnipeg, the city on the flood-prone confluence, as its capital.

If it is ever accepted by the International Commission on Stratigraphy, the already culturally secure Anthropocene concept would officially become the third epoch of the Quaternary period, marking the slip from a history that unfolds *in* geological timespace to a history that *creates* geological timespace.[16] In this stratigraphic mode of reading and writing time, climate change is an anthropogenic extension and intensification of the last glacial retreat, one that will be inscribed into the planet's existing surface, one that has arisen in the sphere of unintentional agencies. Like scrapings of glacial sedimentary materials leaving evidence of their transport, humans intentionally etch their earthworks into the crust, also leaving a myriad of unintentional signs of passing.

At this point, I shift this continental scale story into Manitoban territory in two geological versions: a nineteenth-century tale of glacial advance and retreat and a twenty-first-century version of threshold-crossing, the four-part outburst.

A nineteenth-century Winnipeg origin story, beginning just before water's power assumes the form of great frozen lobes pushing down from the pole as the Laurentian Ice Sheet (paraphrased from Upham 1895):

> Before the last ice age, braided river basins carve out a trough through the continental heart, forming a foundation for the flood bowl now inhabited by Winnipeggers. When Pleistocene glaciers mass at the North Pole, they encounter bedrock as they expand south and downward into this trough. Bedrock guides moving glaciers even as glaciers break up bedrock. Ice plows up bedrock surfaces and carries along its fragments as boulders and gravel. In the ice-rock encounter, glaciers transform bedrock's geological identity, turning solid masses into "drift." Entrained in the bottom of moving glaciers, drift writes upon new bedrock over which it scrapes, leaving scratches and marks called striae.

Bedrock turns into writing instruments and surfaces for writing upon, communicating with scientists across many millennia. Ice/waters' unintentionality meets human intentionality in the guise of geologist Upham and his assistant, Mr. Young. They travel with horse and wagon, reading the striae in the Red River valley. They travel across what had just become the Canadian province of Manitoba (although they didn't seem to be entirely aware of the colony's independence from Britain), orienting data collection in reference to the new transcontinental railroad. Interpreting changes in vanished aquatic forms of geological time, Upham writes for the new field of geology. The striae patterns provide evidence that North American glaciers move along a north-south gradient. The finding contrasts with the east-west gradient Agassiz's European teams find, which contrast indicates how places lend diverse shapes to ice/waters' forces (paraphrased from Upham 1895, 108).

> As earth warms, bringing the Pleistocene epoch to its end, glaciers degrade, forming braided rivers that, voluminously enhanced by all the melted ice, conquer the continental surface once again. But this time, as rivers move north toward Hudson Bay, they find no escape. Even in retreat, the Laurentian Ice Sheet holds its power. Resisting incoherence, the retreating ice sheet blocks the way. The ice dam captures flow from its own meltwaters. Spilling out from different portions of retreating glaciers, trapped rivers come together to form lakes. The edges of Lake Agassiz, the biggest lake, lap against the ice dam that blocks Hudson Bay. Four millennia pass with meltwaters trapped and circulating in the flood bowl's fully submerged glacial transitionary form. Meanwhile, under Lake Agassiz's bed, running straight north, a stream of water digs a shallow channel: the origin of the Red River.

Relationships and identities of the geological past persist in Winnipeg's present. Winnipeg is founded on the confluence of the Red River and the west-to-east-flowing Assiniboine River. Together, the two rivers organize city and surrounding farms, shaping possibilities and limits of flooding and flood control, and ultimately, the viability of Winnipeg as a port. Together, they organize all the city's infrastructure-enabled movement. Although the Red and Assiniboine Rivers have lost the central transport role they played in the time of canoes and York boats used by First Nations and first waves of settlers, the intersection of their meandering channels and flood plains creates the design of local and global transport. Roads and railroads weave around their meandering shapes, around and through neighborhoods that emanate like quilt pieces from their confluence, erroneously but traditionally called The Forks, a meeting point for inhabitants and travelers (figure 2.2).

The Demise of Lake Agassiz in Four Acts: A twenty-first-century Winnipeg origin story (from Teller, Leverington, and Mann 2002 and citations therein):

Two centuries later, still working on Manitoban events, geoscientists collectively produce a radically different kind of imaginative leap into this origin story. They utilize the most sophisticated technological advances in visualizing Quaternary events. Together with quantitative modeling techniques that extend the range and nuance of sensible measurement, geoscientists capture the glacial undoing that forcefully moves abundant freshwater from continental interior to ocean surrounds. These events, they hypothesize with confidence, probably change the course of the oceans' thermohaline currents into the very same routes that sea captains steer their ships upon today.[17]

> By the time the ice dam blocking Hudson Bay collapsed, triggering the largest outburst in the last ice age, Lake Agassiz had merged with glacial Lake Ojibway, to form a superlake with a surface area of about 841,000 square kilometers, more than twice that of the Caspian Sea, the modern world's largest lake, and larger than modern-day Hudson Bay. Lake Agassiz-Ojibway is the only lake in North America to have abruptly released huge volumes of stored water. Outbursts cut through what are now the Mississippi River Valley, St. Laurence River Valley, Mackenzie River Valley, and Hudson Strait. Following the initial outbursts, baseline flow would resume along the new routes to the ocean (see bold tentacular arrows in figure 2.3). Meltwater from these outbursts joins the oceans in a very short time period, "in critical locations and at optimal stages in the evolution of ocean circulation." Combined with longer-term fluxes of Lake Agassiz, the outbursts "probably played an integral role in global ocean and climate history during the last deglaciation," *pushing the system over a threshold.* (paraphrased from Teller, Leverington, and Mann 2002, 883–85)

Teller, Leverington, and Mann theorize that any one of four episodes in which large volumes of lake water suddenly burst out of the confining but collapsing walls of the Laurentide Ice Sheet may have altered circulation patterns of ocean currents and triggered a series of widespread climate changes. Ice/water's logistical power at continental scale breaks into planetary scale when it pushes the earth's

FIGURE 2.2. Red River and railcars moving toward the Forks. View from Canadian Museum of Human Rights. Photo by author.

system over a threshold. Crossing that threshold catapults relations between freshwater circulating in glaciers and rivers and saltwater in oceans into the Holocene configuration that humans find when they arrive in North America. In the map-image (figure 2.3(A)), Winnipeg, though unmarked, is centrally located near what was once Lake Agassiz-Ojibway, well within the source terrain from which the global impulse projects itself toward the oceans.

The authors' quantitative model and accompanying narrative briefly sketched here characterizes unfolding events in geological timespace. This diachronous representation contrasts with their map-image, which collapses into a unified and flattened form the hypothetical events that unfold over several millennia. As a cultural artifact, the map-image effectively projects the viewers' mind into the condensed past of deep time. In addition to condensing time, the map-image contains a cartographic sleight of hand: the ghostly map lines of sovereign nation-states penned lightly into the background insert a geo-politics that is, in fact, not even a glint in the planet's eye.

If the planet crosses over yet another threshold in this century, we shall catapult out of the Holocene and into the unknown. What humans do now with scientific knowledge will—in the best of scenarios—allow the planet to stabilize in a new, *gradually* warming interglacial state. In the worst scenario, it could force the nightmare trajectory toward a "Hothouse Earth" (Steffen et al. 2018, 8253, their figure 1), destabilizing life on the planet.

On the wings of the Anthropocene concept, geoscience pushes out of expert spaces and into broader cultural ones. Human collectives are coming to understand the significance of events taking place in geological spheres of unintended human and aquatic agencies. With nuanced attention to human-ice/water relations, humanities scholarship can craft ethical and political questions, experiments, and stories to help navigate unfolding planetary processes. Here follows one last retelling, this one integrating deep past, present, and future timespaces.

ICE/WATERS' AGENCIES AND WINNIPEG'S ANTICIPATORY PORT (PATH DEPENDENCIES)

Steel Routes of Transnational Capitalism

For better or worse, CentrePort investment speaks to the adaptability of infrastructural logistics. Old pathway dependencies may fade as a melting Arctic lures global trade to the top of the world.[18] The plans of Winnipeg's port boosters, if successful, would confound a basic assumption about port cities. Throughout most of human history, port cities have drawn logistical power from their spatial location in strategic coastal and/or riverine edges. In these geographic situations, they are poised to link water routes and land markets. However, if the future unfolds as the global shipping industry and Canadian maritime government officials anticipate, CentrePort would become an "Arctic Bridge" between Canada and Russia, and between Atlantic and Pacific Oceans (Arctic Council 2009, 11–12). Setting at the northernmost end of the once "NAFTA Highway" (since 2018, the "Mid-Continent Trade Corridor"), cargo would move through CentrePort to and from Russian and Asian markets by three Class I railways,[19] a 24/7 global air cargo airport, and an international trucking hub (see cities along railroad tracks in figure 2.3).[20]

Conventional geovisualizations of port cities rely on images of ships crossing oceans heading back and forth between preset latitude-longitude points along the edges of continental or island land masses. For ports that don't yet have conditions to thrive, that is, for spatialized investment regimes that anticipate profits from future conditions, this visual convention must be reimagined. CentrePort boasts twenty thousand acres on the city's outskirts ready to host port-related investment. Geovisualizations supplement this landlocked footprint with multimodal infrastructural extensions (ship, train, truck networks). These intensify use of existing path dependencies (e.g., railroad lines linking the Red River basin to those in the Mississippi River basin) by exploiting the possibility of future path dependencies across the Arctic Circle. In this way, the Gateways Map illustrates the logistical power of port cities to concentrate and interconnect anticipated nodes of multimodal transport activity dependent on ice-free Arctic passage.

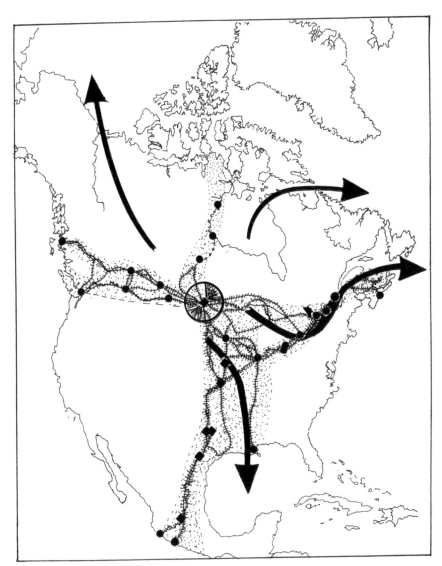

FIGURE 2.3. The Double Artifact: A) geoscience and B) port city projection. A) The four bold tentacles represent the Lake Agassiz outburst (geovisual data from Teller, Leverington and Mann 2002: 880). B) CentrePort's Gateways map of railroads (hatched lines) and cities (circles and squares), Winnipeg in center circle. The singular artifacts overlie NAFTA's ghostly cartography. Original composite drawing by author, from book manuscript (in prep) "Geo-culture: The Ethnography of River-City Flood Control." Composite image by author.

The Gateways' graphic is a cartographic image; an industrial, maritime artifact made to impress potential investors. It projects the viewer's mind into a potential future when CentrePort routinely brings places like Murmansk, Russia, in the

Arctic Circle, together with New Orleans through Churchill, the Hudson Bay port, and Winnipeg, creating new commodity connections among the world's biggest markets. Note that the image-map space between Winnipeg and Churchill, absent a railroad line, remains sketchy. Nevertheless, the website's optimistic text elaborates on developing connections among new and older transport infrastructures. The logistics of future success rely on a graphic assumption that shipping will shift northward as twenty-first-century climate change occurs.

Back to the Spheres of Unintended Agencies: A Visual Experiment

In North America, the logistical power of fresh water derives from ice. This is the lesson I learn from the glacially inspired imagination of Teller, Leverington, and Mann's outburst model. The struggle for power between glaciers and meltwaters and the eventual stabilization of ice/water interfaces and ocean currents are significant geological events. Together, liquid and solid, fresh and salt waters contributed to moving the planet across the epochal boundary between Pleistocene and Holocene. When glacial lobes of the Laurentide Ice Sheet pull back to the pole, they free meltwaters coming from their own bodies. Yet its meltwaters circulate, still, in the very patterns born of the ice sheet's partial defeat. The port city nodal organization of railroad, road, ship, and air lines of twenty-first-century transport grows directly into and around the seasonally recursive versions of these flows and impasses.

I discovered the degree to which the linear forces of human transport correspond to those created by the outbursts from Lake Agassiz quite unintentionally. The discovery emerged as part of my larger geo-cultural experiment. Working my way through various archival pathways, I came upon a telling visual coincidence that illuminates how geological spheres of unintended agencies govern CentrePort aspirations.

The graphic coincidence appearing in my data juxtaposes two timescapes (see Adam 1998). Each is situated in a different layer and portion of geological time. Looking into the past, the outburst map presents a current scientific model of North American flows at Pleistocene's end. The graphic collapses several millennia of ice/water history in which the topography of the continental crust and pathways of the ocean currents come to assume the forms and positions familiar today, whereas, looking into the future, CentrePort's Gateway map presents a portion of the current industrial global transport imaginary that feeds on an open Arctic. The CentrePort graphic collapses timescapes between the nineteenth- and twenty-first-century investment horizons, including those in which humans construct nation-state territories and build the interconnecting railroads, as well as Winnipeg's current investor dream of becoming a global port.

One looks to the past, one looks to the future, but *both occupy the same space.* I study the two heretofore unrelated artifacts: each depicts a global impulse that connects the geographic center of North America with three oceans (Gulf of Mexico, Arctic, and North Atlantic). In each, the global impulse is rendered in the form of powerful and expansive tentacles projecting outward from Winnipeg and its environs.

The coincidence emerges because of my ethnographic impulse to experiment with different kinds of ice/water knowledge, juxtaposing ideas and things that don't usually belong together. And the coincidence extends beyond the intertwining outburst-rivers and railroads. In both artifacts, I can see faint outlines of sovereign territories that situate the emerging tentacles.[21] Drawn with the thinnest of pens, the continental and island coastlines, the divisions between countries (and in one, states and provinces) appear clear yet ghostly beneath the dramas of spectacular global expansion. The repetition of framing is odd given that happenstance brings about the artifacts' relationship: the artist-cartographers use almost the exact same base map. So similar are they, that with a little adjustment of scale on the photocopier, I easily combine the images into one without distorting the content or frame of either. The experiment inspires a visual analysis of intertwining and emanating flows of water, steel, and knowledge.

There is something significant in the trick of global scale that renders space inhuman in the doubled artifact from geoscience and trade. The significant something arises from a humanities perspective on ice/water. In both, the sleight-of-hand plane formed by the pale outlines of sovereign territories set off a viewer's sense of sheer magnitude and irrepressible directionality of the tentacles projecting across abstract land-ice-sea space. The renderings of global impulses need the barest reference to the world we know so that viewers can orient and sense proportion. A vague but familiar geopolitical map underpins a theory of geoscientists and the anticipation of global shippers. The maneuver enables the artifacts as effective visual technologies, essentially making distant past and future seem plausible.

Together, map-images in the double artifact hold the codes of an empirically grounded imaginary that stretches from ice age deep time to transnational capitalist futures. I realize that by combining them, I read things into them that their creators do not intend. For me, the coincidence in the visual rendering of global impulses provides sufficient justification for this ethnographic experiment that takes seriously the spheres of unintended agencies. An industrial dreamland enabled by empirically grounded yet hypothetical geoscientific ghost-outbursts offers a dynamic for unpacking human-water relationships across environmental scales. The double artifact sidesteps the often-tedious boxes of ecological habitats with stakeholders nested neatly in watersheds, nations, regions, and policies. And like the Arctic itself, it opens pivotal geological timespaces to cultural exploration.

CONCLUSION: ICE ENLIGHTENMENT

Our ability to create conditions that may knock the planet out of the Holocene, the epoch that has been so kind to the development of our species and our more-than-human cohabitants, has been discovered only after the fact, or at least, after much harm has been done. This ability to do great harm was unintentional. So even as we must hope for, energize, and rely upon the good and smart intentions of humans to act upon the knowledge signaled by the geo-cultural communiqué of the Anthropocene, we cannot be so arrogant as to assume that the ability of humans to intend is sufficient. Even with ever-advancing technology, the reach, and certainly the nuance, of the sparking but ever so brief presence of human intentionality is no match for the complex scope of planetary life processes. We have to make an ally of ice, understanding its shape-shifting formations and qualities and, most importantly, working with ice's logistical powers of stabilization.

This idea returns to Chandra Mukerji's (2009 and this volume) insight that water lends its agency to people in the form of logistical power. To get water to flow through the Canal du Midi from Mediterranean to Atlantic, engineers, surveyors, and local peasants with skills inherited from the Romans had to change relations among rocks, soils, sands, mountains, and a monarchy. In the exchange of agency, water changed relations well beyond their intentions. In seventeenth-century France, nineteenth-century Erie in New York, Suez in Egypt, or early-twentieth-century Panama, if engineered correctly, water flows, connecting oceans in unnatural ways and changing destinies of port cities and nation-states. And in the twenty-first-century Arctic, as ice diminishes with warming temperatures, water flows into channels cracked open by ice-breaking ships in the Canadian Archipelago and Siberia. In a process of what Mukerji might call "impersonal rule" although involving multinationals rather than monarchies, ice-breaking ships are apt to destroy or confuse time-honored trails of Indigenous hunters and animals.[22] Old and new, trails, canals, and channels crisscross the earth, material evidence that human collectivities, like glacial lobes, inscribe movement, stasis, struggle, and art into the always revising surface of Earth (figure 2.4).

The spatial coincidence that emerged from combining geoscience and port industry geovisualizations shows that if the well-frozen Arctic to its north disappears, Winnipeg might bear the fruits of a multimodal system that links Siberia to the Gulf of Mexico. Drawing on the logistical power of ice/water, the city is planning to extend land and air tentacles into some of the new-Holocene paths carved out by Lake Agassiz-Ojibway's original outbursts. Indeed, if CentrePort boosters are right, the city's near-term future might shine as brightly as it did in its Golden Age when railroads first came through town. But it would shine quite briefly, I fear. In the longer term, crossing the next climate threshold (a possible event hastened by a melting Arctic) would not be survivable. There is no coming back from the bust that would almost surely follow this boom. Is there a way to hold such promise and threat together in one's mind? Might it be possible for Winnipeggers and global shippers more generally to borrow meltwaters' logistical power without hastening a more dangerous future?

Such questions can be put to work. Humanities scholars experimenting with ways of knowing and being can wield revelatory coincidence, loosening the

FIGURE 2.4. Ice Sculpture. When the air is cold enough and the ice is deep enough, artists sculpt the Red River at the Forks. February 10, 2007. Photo permission granted by Beverly Peters, Winnipeg.

fixities of our intentions to clarify the diversity of our options. By "staying with the trouble," as Donna Haraway (2016) suggests, this chapter writes a city into glacial timespace, joining others in a call to protect the ice/waters that move us.

ACKNOWLEDGMENTS

The material presented here draws from a book-in-progress provisionally entitled "Geo-Culture: The Ethnography of River-City Flood Control." I received support for the 2014 field research as Resident Chair in Environmental Science at University of Winnipeg's Richardson College for the Environment and Science, from the Council for International Exchange of Scholars (CIES), Fulbright Canada, and the College of Arts and Sciences and Vice Provost for Research at Indiana University. I presented an earlier version at the annual meeting of the American Anthropological Association in Washington, DC, on December 2, 2017. More generally, the ideas have been inspired by ongoing conversations about Arctic ice and global shipping with Claudio Aporta, Aldo Chircop, Kate Coddington, and Phil Steinberg and by participation in the ICE LAW Project, supported by the Leverhulme Trust (Grant IN-2015–033). I thank the volume's editors for their queries and comments along the way.

REFERENCES

Adam, Barbara. 1998. *Timescapes of Modernity: The Environment and Invisible Hazards*. New York: Routledge.

Arctic Council. 2009. *Arctic Marine Shipping Assessment 2009 Report*. https://oaarchive .arctic-council.org/handle/11374/54, accessed December 10, 2018.

Aporta, Claudio. 2009. "The Trail as Home: Inuit and Their Pan-Arctic Network of Routes." *Human Ecology* 37 (2): 131–46.

Aporta, Claudio, Stephanie C. Kane, and Aldo Chircop. "Shipping Corridors through the Inuit Homeland." *Limn* #10. https://limn.it/articles/shipping-corridors-through-the-inuit -homeland/.

Carse, Ashley. 2014. *Beyond the Big Ditch: Politics, Ecology, and Infrastructure at the Panama Canal*. Cambridge, MA: MIT Press.

Carter, Natalie, Jackie Dawson, Jenna Joyce, and Annika Ogilivie. 2017. *Arctic Corridors and Northern Voices: Governing Marine Transportation in the Canadian Arctic*. Environment, Society and Policy Group, University of Ottawa. www.espg.ca.

Chakrabarty, Dipesh. 2009. "The Climate of History: Four Theses." *Critical Inquiry* 35 (2): 197–222.

Chircop, Aldo. 2016. "Sustainable Arctic Shipping: Are Current International Rules for Polar Shipping Sufficient?" *Journal of Ocean Technology* 11 (3): 39–51.

Cowen, Deborah. 2014. *The Deadly Life of Logistics Mapping: Violence in Global Trade*. Minneapolis: University of Minnesota Press.

Desfor, Gene, Jennefer Laidley, Quentin Stevens, and Dirk Schubert, eds. 2011. *Transforming Urban Waterfronts: Fixity and Flow*. New York: Routledge.

Dodds, Klaus, and Mark Nuttall. 2016. *The Scramble for the Poles*. Malden, MA: Polity.

Ghosh, Amitav. 2016. *The Great Derangement: Climate Change and the Unthinkable*. Chicago: University of Chicago Press.

Hansen, Bert. 1970. "The Early History of Glacial Theory in British Geology." *Journal of Glaciology* 9 (55): 135–41.

Haraway, Donna. *Staying with the Trouble: Making Kin in the Chthulucene*. Durham, NC: Duke University Press.

Irmscher, Cristopher. 2013. *Louis Agassiz*. Boston: Houghton Mifflin Harcourt.

Kane, Stephanie C. 2012. *Where Rivers Meet the Sea: The Political Ecology of Water*. Philadelphia: Temple University Press.

———. 2018. "Where Sheets of Water Intersect: Infrastructural Logistics and Sensibilities in Winnipeg, Manitoba." In *Territory beyond Terra*, edited by Kimberley Peters, Philip Steinberg, and Elaine Stratford, 107–26. London: Rowman and Littlefield.

Kelly, Jason M., Philip Scarpino, Helen Berry, James Syvitski, and Michel Meybeck, eds. 2018. *Rivers of the Anthropocene*. Oakland: University of California Press.

Kives, Bartley, and Bryan Scott. 2013. *Stuck in the Middle: Dissenting Views of Winnipeg*. Winnipeg: Great Plains Publications.

Krupnik, Igor, Claudio Aporta, Shari Gearheard, Gita J. Laidler, and Lene Kielsen Holm. 2010. *Siku: Knowing Our Ice; Documenting Inuit Sea-Ice Knowledge and Use*. New York: Springer.

Levinson, Marc. 2006. *The Box: How the Shipping Container Made the World Smaller and the World Economy Bigger*. Princeton NJ: Princeton.

Linton, Jamie. 2010. *What Is Water? The History of an Abstraction.* Vancouver: UBC Press.

Mah, Alice. 2014. *Port Cities and Global Legacies: Urban Identity, Waterfront Work, and Radicalism.* New York: Palgrave Macmillan.

Mezzadra, Sandro, and Brett Neilson. 2019. *The Politics of Operations: Excavating Contemporary Capitalism.* Durham, NC: Duke University Press.

Mukerji, Chandra. 2009. *Impossible Engineering: Technology and Territoriality on the Canal du Midi.* Princeton: Princeton University Press.

Raffles, Hugh. 2002. *In Amazonia: A Natural History.* Princeton: Princeton University Press.

Schmidt, Jeremy. 2017. *Water: Abundance, Scarcity, and Security in the Age of Humanity.* New York: NYU Press.

Schneiderman, Jill S. 2015. "Naming the Anthropocene." *philoSOPHIA* 5 (2): 179–201.

Sideris, Lisa. 2016. "Anthropocene Convergences: A Report from the Field." *In Whose Anthropocene? Revisiting Dipesh Chakrabarty's "Four Theses,"* edited by Robert Emmett and Thomas Lekan. *Transformations in Environment and Society* 2: 89–96.

Somerville, Margaret. 2013. *Water in a Dry Land: Place-Learning through Art and Story.* New York: Routledge.

Steffen, Will, John Rockström, Katherine Richardson, Timothy M. Lenton, Carl Folke, Diana Liverman, Colin P. Summerhayes, et al. 2018. "Trajectories of the Earth System in the Anthropocene." *Proceedings of the Academy of National Sciences (PNAS)* 115 (33): 8252–59.

Steinberg, Philip E. 2001. *The Social Construction of the Ocean.* Cambridge: Cambridge University Press.

Steinberg, Philip, Greta Ferloni, Claudio Aporta, Gavin Bridge, Aldo Chircop, Kate Coddington, Stuart Elden, Stephanie C. Kane, et al. 2022 (forthcoming). "Navigating the Structural Coherence of Sea Ice" (working title). In *Unsettling Ocean Legalities,* edited by Irus Braverman. Oxfordshire: Routledge.

Strang, Veronica. 2004. *The Meaning of Water.* Oxford: Berg.

Subramanian, Meera. 2019. "Anthropocene Now: Influential Panel Votes to Recognize Earth's New Epoch." *Nature.* May 21, 2019. www.nature.com/articles/d41586-019-01641-5.

Syvitski, Jaia, Colin N. Waters, John Day, John D. Milliman, Colin Summerhayes, Will Steffen, Jan Zalasiewicz, et al. 2020. "Extraordinary Human Energy Consumptions and Resultant Geological Impacts Beginning Around 1950 CE Initiated the Proposed Anthropocene Epoch." *Communications Earth and Environment* 1(32): 1–13.

Teller, James T., David W. Leverington, and Jason D. Mann. 2002. "Freshwater Outbursts to the Oceans from Glacial Lake Agassiz and Their Role in Climate Change during the Last Deglaciation." *Quaternary Science Review* 21: 879–87.

Tsing, Anna Lowenhaupt. 2005. *Friction: An Ethnography of Global Connection.* Princeton: Princeton University Press.

Upham, Warren. 1895. *The Glacial Lake Agassiz.* Washington, DC: U.S. Geological Survey.

Whyte, Kyle P. 2018. "Indigenous Science (Fiction) for the Anthropocene: Ancestral Dystopias and Fantasies of Climate Change Crises." *Environment and Planning E: Nature and Space.* 1 (1–2): 224–42.

Williams, Mark, Jan Zalasiewicz, Neil Davies, Ilaria Mazzini, Jean-Philippe Goiran, and Stephanie Kane. 2015. "Humans as the Third Evolutionary Stage of Biosphere Engineering of Rivers." *Anthropocene* 7: 57–63.

Yusoff, Kathryn. 2016. "Anthropogenesis: Origins and Endings in Anthropocene." *Theory, Culture, & Society* 33 (2): 3–28.

NOTES

1. This chapter is part of a larger ethnographic field-based book project on flood control in Winnipeg Canada. My working title is "Geo-Culture: The Ethnography of River-City Flood Control." Figure 2 comes from the manuscript (under review).

2. For a recent statement on the question of onset and official geological status of the Anthropocene, when human action triggered climate change and its widespread effects, see Subramanian (2019) and Syvitsky et al. (2020).

3. Ethnographers also tend to import the basic hydrological assumption that water is composed of H_2O molecules. Not an unreasonable assumption perhaps. As they gather and flow, cycle, and change state with temperature, gravity, and wind, H_2O molecules implicitly provide elemental material for cultural meaning. When oceans and rivers become performance stages for human action (e.g., ships plying the sea) or habitats for human and more-than-human relationships (e.g., petrochemical and plastic pollution), their molecular composition as H_2O (plus dissolved compounds) is taken for granted. But it is also possible that water's atomic character somehow conceals embodied senses of water's agency. The extent to which water's molecular nature governs meaning in hydrological science is the subject of Jamie Linton's (2010) "history of a modern abstraction." He analyzes the conceptual limitations of water stripped of its environmental, social, and cultural contexts. And also, when the idea of water/H_2O turns into a "total geological fact," as Jeremy Schmidt (2017, 4, 14, 208) finds in his intellectual history, the water's meaning is reduced to its stature as an absolutely necessary resource. In turn, water managers promote themselves as practitioners who make social life possible and as key agents of modern planetary progress.

4. The ethnography of water is extensive; for a range of approaches see, for example, Raffles (2002), Strang (2004), Kane (2012), and Somerville (2013).

5. I use a definition of agency inherited from seventeenth-century English: "Ability or capacity to act or exert power; active working or operation; action, activity." *Oxford English Dictionary* (2:4).

6. This outward exploration of unintentional spheres reverses the inward orientation of psychoanalysis.

7. With Jaia Syvitski and Kimberly Rogers of CSDMS, University of Colorado, I have been exploring meso-scale spaces of water infrastructure in a project titled "Humanizing Remotely Sensed Visions of the Earth: River Infrastructure in Image, Ethnohistory and Geohydrology."

8. I learned about CentrePort when I went to interview its vice president for planning and development in its downtown office to discuss flood control (the focus of my larger ethnographic project). Interview with John Spacek, November 21, 2014, field notes book no. 4, 64–75.

9. I first developed the concept of *infrastructural culture* to capture the broad cultural significance of flood control engineering practices in Winnipeg (Kane 2018). I then broadened the concept and renamed it as *geo-culture* to refer to the way urban inhabitants embed themselves in the crust of the earth and, not so unlike the glaciers of the last ice age, reshape it. By approaching human action through an ethnography of cities and their situated intra-actions with ice/water bodies, I sidestep much of the "homogeneous geomorphizing of the Anthropocene" (Yusoff 2015, 3) that accompanies critical debates of species-level analysis (e.g., Chakrabarty 2009; Sideris 2016; Whyte 2018).

10. The IMO is the United Nations agency responsible for the safety and security of world shipping and for protecting the seas and atmosphere from ship pollution. www.imo.org/en/About/Pages /Default.aspx, accessed December 10, 2018. Note too that between 2015 and 2017, shipping traffic has increased 75 percent, mostly in Nunavut Province, homeland of the Inuit and directly north of Winnipeg's provincial home of Manitoba (Carter et al. 2017, 4).

11. www.npolar.no/en/facts/albedo-effect.html, accessed December 10, 2018.

12. For more on the interdisciplinary research network called the Ice Law Project, convened by the Centre for Borders Research at Durham University, 2014–19, see https://icelawproject.org/.

13. See, for example, Aporta (2009); Aporta, Kane, and Chircop (2018).

14. The Ice Law Project (ILP) investigated the potential for an ice-sensitive legal framework for governing potentially conflicting activities such as global shipping and Indigenous livelihoods in the frozen Arctic Ocean. The ILP archive is available at: https://icelawproject.weebly.com/.

15. Phil Steinberg clarified Chircop's argument for me in an email, February 16, 2019. Aldo Chircop raised the legal problem in a coleader discussion of the Ice Law Project in Stockholm with Phil Steinberg, Stuart Elden, and me on June 21, 2017. See also Steinberg et al. (2022, forthcoming).

16. For rivers as key actors in the scholarly materialization of the Anthropocene, see Williams et al. (2015) and Kelly et al. (2018).

17. National Ocean Service, Currents: "Thermohaline Circulation." https://oceanservice.noaa.gov/education/tutorial_currents/05conveyor1.html, accessed November 20, 2018.

18. See also: www.centreportcanada.ca.

19. Canadian National, Canadian Pacific, and BNSF Railways.

20. See note 18.

21. The geoscience article also shows latitude (30 and 60 degrees) and longitude (30 to 150 degrees). Both geoscience and trade images extend partway into the Arctic and partway into Mexico.

22. See, for example, www.ntkp.ca/ and www.anijaarniq.com/.

3

Radical Water

Irene J. Klaver

Water is everywhere before it is somewhere. . . . It is a terrain that challenges assumptions, reminds us of our fallibility, accommodates complexity, and locates our horizon. (Anuradha Mathur and Dilip da Cunha 2014, x–xi)

"There it is—take it!" With this legendary short dedication speech, William Mulholland, superintendent of the Los Angeles Water Company, opened the Los Angeles Aqueduct in 1913 (Mulholland 2000, 246). As chief engineer, he envisioned and supervised the project to bring water from the Owens River Valley across 220 miles of rough, mainly desert terrain to Los Angeles through an elaborate system of canals, syphons, tunnels, and pipes. More than one hundred years later in 2020, on the other coast, New York City mayor Bill de Blasio warned citizens of the potential effects of Tropical Storm Isaias: "Take this very seriously" (Schuman 2020). This seems to be a big leap forward. The mayor's warning recognizes that water is not just a passive substance to be funneled into human projects. It recognizes that water has power: the power of deadly flooding, the capacity for wreaking havoc on human structures. The mayor's call can be seen as a call to war on water. Water is the enemy, the Other. We need to arm ourselves and fortify the city: the water will come.

Here I argue for another way of taking water seriously: a relational way. This perspective acknowledges that water is always in relation; it is not the enemy, the Other. Water has become the enemy because of our own engineering designs for controlling it and separating it from us, from the land. To acknowledge water as intrinsically relational opens a different sense of water and a different water. It entails a shift from modern water to relational water. In this shift it is no longer water as such that causes floods and problems, but *modern* water that is at fault. I show how water, in engendering this move, has agency. I argue that water, therefore, is radical.

Jamie Linton coined the term *modern water*. The term elucidates how water in the modern era (from the Scientific Revolution in the seventeenth century and the eighteenth-century Enlightenment until now) has become homogenized and universalized into H_2O and how this reduction came to be understood as water's true essence, its basic nature. It denudes water of its ecological, cultural, social, and political dimensions and relations, making it easier to manage (Linton 2010, 7–8). Maria Kaika points at the "productivist instrumental rationality that came to permeate all facets of modern life," fueled by a strong belief in "human emancipation through the domination of nature." This nature/society dualism came with "a fragmentation of everyday experience, and the increasing commodification of everyday life" (Kaika 2005, 12–13). Modern water, "as the dominant, or natural way of knowing and relating to water, originating in Western Europe and North America," had come to operate "on a global scale by the later part of the twentieth century" (Linton 2010, 14). Linton argues that the so-called crises of water scarcity, of water pollution, are not crises of water per se, but crises of *modern* water: "modern water itself establishes the epistemological conditions that inevitably give rise to crisis" (23). He calls for the adoption of more flexible hydrosocial relations. It is the "relation that defines the essence of what water is" (223). In this primacy of relation, knowing water is a product of engagement. Radical water is relational.

In "Indigenous Peoples and the Politics of Water," Melanie K. Yazzie and Cutcha Risling Baldy develop the notion of *radical relationality* (Yazzie & Baldy 2018, 2). It is a term "that brings together the multiple strands of materiality, kinship, corporeality, affect, land/body connection, and multidimensional connectivity coming primarily from Indigenous feminists.... It provides a vision of relationality and collective political organization that is deeply intersectional" (2). Relational water is no longer a resource, but a relative "with whom we engage in social (and political) relations premised on interdependence and respect" (3). Fostering radical relationality is part of a collective, cooperative struggle to build "vibrant alternative futures.... How we struggle is how we remember, how we live . . . It is how we *relate*. This is what water teaches us" (12). Leanne Betasamosake Simpson in *As We Have Always Done: Indigenous Freedom through Radical Resistance* (2017) emphasizes the central place of storytelling in imagining radical futures. Radical resurgence happens through the radical diversity of everyday stories.

Modern water, homogenized and contained, is cut off from stories, from relations. It means other entities have also been cut off, contained, and homogenized. Swamps and wetlands are drained, aquifers pumped dry, rivers diverted—to be used for agriculture, for development. Water's isolation, water's separation from relation, makes water bodies measurable, static, determinate, facilitating domination, exploitation, commodification, and colonization of water and of everything with which it was in relation. Furthermore, by separating water from relation, making it modern, a line is drawn which makes water the Other; in crossing this line, modern water creates the notion of flood. Relationality, on the other hand, entails absence of hard boundaries; its intrinsic diversity displays fuzzy boundaries of indeterminacy and complexity and, therefore, is harder to control. There is unpredictability and uncertainty. Contained and restrained water becomes modern water, global water, a commodity, an asset performing increasingly well on stock markets.

Contained, diverted, dammed, homogenized, cut off from its relations, water snaps, as in Sara Ahmed's (2017) "feminist snap," a "collective snap," not "a single moment of one woman experiencing something as too much," but "a series of accumulated gestures that connect women over time and space" (200). With many different bodies of water impacted, water's "series of accumulated gestures" includes a range of manifestations. The disappearance of springs, the shrinking of small lakes, the rising of the sea level, which pushes the groundwater up and makes pollutants, stored in the soil for years, for decades, suddenly surface, bleeding in the air, in the water. Stratified rivers keep flooding houses built all the way to the waterfront, washing away cars, streets, vegetation, things, animals, humans. Melting glaciers crashing into dams leave small villages covered in mud floods laced with debris. These are, in Linton's terms, the crises of modern water.

When water snaps, it becomes radical. It refuses to be reduced to simplicity, to the impacts of modernity. Stressed to its limit, water demands radical change—change in how we deal with climate change, with hydroelectricity, with irrigated agriculture, with river "management," with city planning, with building codes, with plastics, with fracking. Radical water demands radical change in our thinking and doing.

Water is radical because it provokes fundamental questions. As Mathur and da Cunha state, water "is a terrain that challenges assumptions, reminds us of our fallibility, accommodates complexity, and locates our horizon" (Mathur and da Cunha 2014, xi). What is not in its terrain?

Fluid and ephemeral, water is the bedrock of the world. Water orients us, shows us how boundaries are interrelated, and not just hard walls; water shows us soft versus hard approaches. It teaches a shift in mentality, in modes of thought, in ways of operating; it teaches us how to live *with* water instead of conquering and dominating it. Underlying this mentality shift, in which water is taken seriously, is a radical incommensurability with the modern conceptualization of water, with how we think we can manage and control "it." Radical water demands a radical overhaul of our conceptualization of water, of our planning and managing water as a separate entity. The incommensurability is on the level of epistemology, ontology, ethics, and aesthetics. It changes what counts as progress, certainty, justice, efficiency. It affects how we conceive of boundaries, time, place, space, relations.

Radical water is multiple. There is not just one way of water. There are many ways of water, and many ways of knowing and experiencing water. Multiplicity and complexity are intrinsic to water. Water is always in relation; it *is* relation. Therefore, it is multiple. Humid, wet, fluid, and frozen, it makes mountains crumble, trees stand straight, people fight and celebrate. Water rhizomes into relations, ramifications, and constellations. It is omnipresent, evanescent, liquescent, ephemeral, multidi-mensional, gestational, conceptual, virtual. Water engages actively and passively; it drips, sits, sinks, mists, dissolves, melts, oozes, flows, freezes, rains, cascades, evap-orates. Because of its relationality, it does not let "itself" be reduced to simplicity, to an incapacity to act. Its "self" is many. Its being *is* becoming. It embodies concepts, rituals, politics, ideas, and ideals. Embodied, it is in other bodies, in other environ-ments, and provides environments for other bodies. Inside and outside, interior and exterior interchange; water is mist, rain, a terrain, mud, microbial, intesti-nal, virtual, and cyborgial. It gives life and takes lives; it can be abundant, scarce, present, absent. It challenges clear-cut divisions and oppositions, undermines categorizations, messes up lines of separation, laughs at institutions, builds and resists infrastructures. It leaks, overflows, erodes, spreads, disappears, dilutes, and pollutes. Being in relation, water is fundamentally indeterminate. Radical water undermines its own categorization as a clear-cut separate entity. It cannot be cut. When it gets cut, it bleeds. When it is confined, it snaps.

Water is complex in its ontological, sociological, political, hydrological, epistemo-logical, religious, cultural, ethical, experiential ways. Water itself shows the above, as we will see below. Water engenders activism and advocacy. Water is prehuman, posthuman, nonhuman. We are not at its center. It is at our center. Water is radical.

In this chapter I look at how we live with water differently by transitioning from modern water to relational water. I look first to "hard" responses to events such as Hurricane Sandy in New York and the beginnings of such responses with the Dutch and nineteenth-century "progress" as lenses for close examination of the colonization of water. I then turn to New Orleans to weave in an Indigenous per-spective of relational water. I argue that colonizing water is already made visible by mere lines on maps. Seeing these lines as instrumentalities of progress distorts our very understanding of water, cutting us off from relational water. I present an integrated understanding of water that could reengage our relationality with water. This mentality shift looks to the notions of sedimentation and reactivation in the philosophy of Merleau-Ponty, as well as to the actual workings of sedimentation and reactivation in the processes of meandering—the dynamic relation between water and land/sediment. Finally, recent thoughts from designer Dilip da Cunha and engineer Klaus Jacob on the relation between water and land and the impact of colonial thinking on water round out my considerations of modern water. I end with how we might liberate water and ourselves by embracing radical relationality, by looking to water to guide us.

MADE LAND

Mulholland's pointed 1913 dedication speech, "There it is, take it," was the inau-guration not just of an aqueduct, but of an era, a new mentality, a new lifestyle, a new mode of water: modern water. Water was to be taken as part of a "trajectory of modernity, of progress by controlling nature for human use" (Klaver and Frith

2014, 520). One hundred years later, New York City was compelled to take the water seriously. In October 2012, Hurricane Sandy hit New York, "flooding more than 88,000 buildings in the city, killing 44 people, and causing over $19 billion in damages and lost economic activity" (Goodell 2017, 147). Sandy was fresh in New York City's memory when eight years later Mayor Bill de Blasio warned NYC citizens about Tropical Storm Isaias: "Take this very seriously." Goodell called Sandy a "transformative event" (147). But to what extent did it change relations with water *radically*?

When the Dutch expanded their colony of New Amsterdam on Manhattan Island in the early seventeenth century, they had already mastered an extensive engineering know-how of windmills used for milling grains and reclaiming land from water (Klaver 2016a). The colony was initially built around Fort Amsterdam (1626), to protect the beaver pelt trade of the Dutch West India Company against looming attacks from other European colonial powers in contestation over the entrance to the Hudson River. The fort functioned as a warehouse for company goods, but also as a safeguard for the settlers' farms and investments. To legally guarantee the safety of the new possessions, the Dutch had "purchased" the island of Manhattan from the Indigenous population in 1625. The new "owners" built sawmills and flour mills, turning the island's many creeks into hydropower. Flour became an important trade good, as did beaver pelts. Flour barrels, beavers, windmills, and the presence of Indigenous and colonizer populations are all symbolized on the seal of the City of New York (figure 3.1).

The City's website describes the two men supporting the shield as follows:

> Supporters: A sailor on the left, his right arm bent, and holding in his right hand a plummet; his left arm bent, his left hand resting on the top of the shield; above his right shoulder, a cross-staff.
>
> A Native American of Manhattan, his right arm bent, his right hand resting on top of the shield, his left hand holding the upper end of a bow, the lower end of which rests on the ground. Shield and supporters rest upon a horizontal laurel branch. (NYC Green Book Highlights n.d.)

The shield shows "the sails of a windmill. Between the sails, in chief a beaver, in base a beaver, and on each flank a flour barrel." By the time the English took over and changed the name to New York in 1665, the colonial water activities of the Dutch had changed the complex water-land relations of the island.

Two hundred years later, New York had become the most important trading port and the largest city in the United States. In 1865, topographical engineer Egbert Ludovicus Viele (1825–1902) captured Manhattan's original water features, such as its shoreline, creeks, underground waterways, springs, and meadows, together with the new sewer lines, in a detailed map, his *Sanitary & Topographical Map of the City and Island of New York Prepared for the Council of Hygiene and Public Health of the Citizens Association* (figure 3.2). Viele had been a military officer in

FIGURE 3.1. The Seal of the City of New York; 2015 colored rendering by K. Lefebvre of Paul Manship's authorized 1915 sculpted version of the seal.

the Mexican War (1846–48) and established a military camp at the Rio Grande close to Laredo, where several of his men died of cholera. Upon his return to civilian life, he had an important role as a civil engineer for New York City's parks. He was in charge of planning Frederick Law Olmsted's concept for Central Park, and he designed Brooklyn's Prospect Park (Segovia 2010). Then, between 1861 and 1864, Viele had various functions in the Union army in the Civil War. He was struck by the fact that the mortality rate from epidemic infectious diseases, such as typhoid, cholera, malaria, and measles, was twice as high as deaths from battle wounds (Sartin 1993, 580). Apparently, "it was said that the suffering he saw, caused by poor sanitation, motivated him to help sewer engineers by mapping the city's

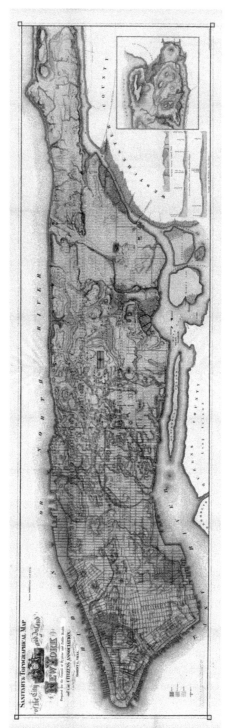

FIGURE 3.2. *Sanitary & Topographical Map of the City and Island of New York Prepared for the Council of Hygiene and Public Health of the Citizens Association.* [U]nder the direction of Egbert L. Viele, Topographical Engineer, 1865.

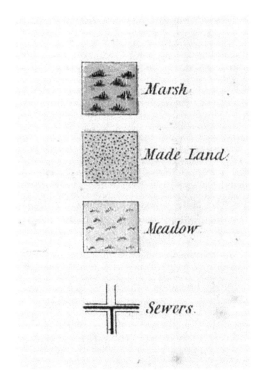

FIGURE 3.3. Legend of the Viele
Map (detail), showing three landforms
and sewer system.

streams" (Kurutz 2006). Cholera was rampant in the mid-nineteenth century, with violent outbreaks in London in the summer of 1854 and in Pittsburgh during the same year. Experience with the lack of sanitation pushed Viele to his meticulous rendering—with military precision—of a new sewage map of Manhattan.

In 1874 Viele had the map republished, lifting it out of the *Report of the Council of Hygiene* and publishing it as *Topographical Map of the City of New York: Showing Original Water Courses and Made Land / Prepared under the direction of Egbert L. Vielé, topographical engineer.* The map showed sewer lines and a street grid superimposed upon three different landforms: Marsh, Made Land, and Meadow (figure 3.3). It revealed how Manhattan was once laced with pools, ponds, creeks, streams, springs, meadows, and marshes. Modern urbanization of the island drained, depleted, and diverted these complex water bodies; it stratified them, paved them over, buried them under high rise buildings, and pushed most of them off the map.

At the same time, the very incorporation of Viele's "original water courses" in a map inscribed them in the modern project. The very act of capturing them as static, precisely lined bodies of water on a map reduced them to modern water. Furthermore, the very fact that Viele republished his map from the *Report of the Council of Hygiene* independently and renamed it *Topographical Map of the City of New York: Showing Original Water Courses and Made Land* indicates that he foresaw that these "original water courses," captured as topographical items, would be of importance in the controlling of New York's water-land relations. And, indeed, his making visible of these already invisible waterways made and still makes Viele's map extremely valuable. It is the only map that shows the old waterways, which is still vital information for civil engineers today in their design of buildings and site developments (Kurutz 2006). If engineers do not take them into account, these water features very likely will become "problems;" they will create a crisis. Reduced to lines instead of breathing water, sinking and soaking in the soil, bubbling in a spring, they are captured and separated. They have become modern water. They have lost their relationality. The very fact that they are captured *on* the map gestures to their fate of being pushed *off* the map.

Another significant aspect of Viele's map is his category of "Made Land." We see how by 1865 the coast of Lower Manhattan had been enlarged with a finely grained, light-brown colored ribbon of "Made Land," surrounding the end of the land arm, tight as a mitten (figure 3.4). When Hurricane Sandy hit New York City in October 2012, most of these areas of Manhattan were under water. The human-made, reclaimed land forms a hard boundary line with water: a line to be crossed, creating a flood. Being lower-lying land, it is more vulnerable to such flooding.

Furthermore, its vulnerability was increased by the human elimination of storm buffers, such as salt marshes and oyster beds. When the water came, there was nothing to slow it down. The water reclaimed its coast.

Hurricane Sandy created new words, new professions, new offices, and new projects. Whether it truly transformed a way of thinking is still to be seen. In 2018, the city's new Office of Resilience and Recovery "planned to break ground on what's called the East Side Coastal Resilience Project, a ten-foot-high steel-and-concrete-reinforced berm . . . the first part of a larger barrier system, known informally as the Big U, that someday may loop around the bottom of Lower Manhattan" (Goodell, 146). The Big U is a typical example of a "hard" approach to water: "a solid wall—a modern rampart against the attacking ocean" (146). Instead of relating to the water, that is, working and living with it, the hard approach fights the water, walls it out. Costing billions of dollars to construct (147), the project shores up efforts to further control and contain water with more—literally "Big"—technology and engineering. As with most infrastructures that divide, the proposed wall building is fraught with politics: "You can't wall off the city's entire 520-mile coastline, so how do you decide who gets to live behind the wall and who doesn't?" (Goodell 2017, 148). Questions such as who is deciding, who will be benefiting from it, and who will be suffering from it are riddled with environmental justice issues, which are often racial issues.

The emblem in the upper-left corner of Viele's map (figure 3.2) reveals how these issues were part and parcel of the beginning of New York, at the beginning of New Amsterdam. The small emblem in the title of Viele's map is a clear reference to the official seal of the City of New York (figures 3.1 and 3.5). Like the official seal, it portrays an Indigenous man and a white man at opposite sides of a shield, which

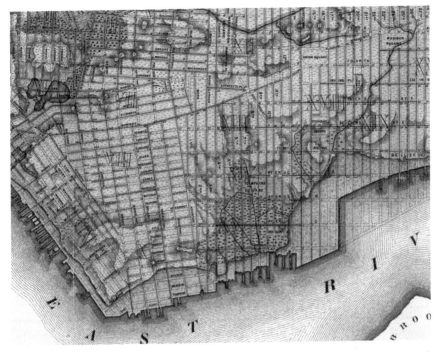

FIGURE 3.4. The Viele Map (detail), showing part of Manhattan's East Village as Meadows and Tompkins Square Park as Marsh, flanked by Made Land.

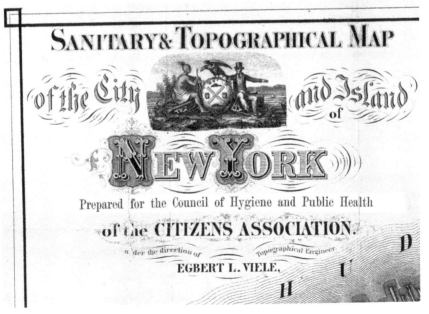

FIGURE 3.5. Title and emblem of the Viele Map (detail).

also depicts a figure X of four windmill sails, again with a beaver in the upper and lower space and a flour barrel in the left and right space. Yet, there are significant differences. The Indigenous man, now at the left side, sits on an animal hide, trees around him in the background, a cloth over his loins, a bow in his right hand, and a quiver with arrows at his right leg. He is no longer looking us in the eye; we see him in a side view: he looks attentively to the white man, seemingly listening to him. The white man is also sitting, looking partially at us, partially at the Indigenous man. He leans against the shield on his right forearm in a relaxed pose, while his left arm stretches out, gesturing over the water, over the horizon, to the wide expanse of land. A sailing ship passes by in the background, indicating more trade. The white man sits on a bale of wheat and has a measuring instrument at his feet. He is dressed in a suit and a top hat. He radiates that he is in charge. He is Man the Maker. He has the dominant position: he measures things, stakes out property, sets up his business, controls nature by subduing it. He is no longer a simple sailor, but a master of the land, the developer, the businessman. The eagle, a symbol of power on the city's seal above both men, seems now allied with the white man, eyeing the Indigenous man in a slightly threatening pose. This scene brings out the colonial relation, more subtly buried in the city's seal, where the sailor and the Indigenous man seem to have a more or less equivalent standing posture. However, the shield—and the sheer presence of a shield as such—reveals how dynamic land and water relations were on their way to being radically changed into disembodied private property lines; how Indigenous peoples and their ways of life were being driven away or eliminated; how waterways became hydropower, used for milling activities; how marshes were pumped dry; how beavers, reduced to pelts, became a valuable asset.[1] Killing beavers killed two birds with one stone by delivering lucrative pelts for trade and by eliminating streams "erratically" engineered by the beavers, keeping waterways under the control of men.

SWAMP LAND

Where the city seal of New York gestures at the transition from Indigenous ways of life to a modern settler colonial occupation of the land, Viele's emblem, and especially his map, show how by the late nineteenth century a modern mindset had become the dominant and dominating power. A long tradition of Indigenous resistance ensued. As Nick Estes writes in his book with the telling title *Our History Is the Future*, "There is one essential reason why Indigenous peoples resist, refuse, and contest US rule: land. In fact, US history is all about land and the transformation of space, fundamentally driven by territorial expansion, the elimination of Indigenous peoples, and white settlement" (2019, 67). Privileging of land is foreign to Indigenous ways of life: land was—and is—intrinsically related to water and all other beings. Estes gives a powerful account of Indigenous resistance at the

Standing Rock Indian Reservation from 2016 to 2017 to block the construction of the Dakota Access oil Pipeline (DAPL), threatening the reservation's water supply. The protesters called themselves Water Protectors; their message: "WATER IS LIFE." They stood "for something greater: the continuation of life on a planet ravaged by capitalism" (15). As Estes asserts with his title *Our History Is the Future*, Indigenous ways of life hold the future. The ways the Indigenous peoples of the large Mississippi delta lived with the complex dynamic water-land relations exemplify Estes's title. They hold the future.

At the mouth of that river delta, Jean Baptiste Le Moyne de Bienville "still roams the streets of New Orleans, the city he platted out of swamp land in 1718" (Howe 1994, 109). LeAnne Howe gives an ironic account of how the "Father of New Orleans" acquired the land for his city:

> My Choctaw ancestors called Bienville, *Filan-chi*, which is short for, "Our Frenchman, The Nail Biter." They liked him, although he was nervous and could never take a joke. The first time they invited him for dinner he started making problems that have continued until now. (109)

Howe tells how dinners in those days were collective events, carefully orchestrated, lasting days, a week, full of jokes, gossip, exchanging information about what was going on where, what other tribal bands were up to, and what the white men were doing. Tone deaf to cross-cultural understandings, Bienville scoffed at the etiquette. He was out to do business. Calculated and efficient, at least in his sense of the word, he tossed "some third-rate glass beads on the ground" before the meal, and asked brusquely "if my relatives would trade them for "*un morceau de terre*" (a morsel of land somewhere) (109). Offended by the insensitive gesture and impertinent question:

> Some wanted to kill him right on the spot; others thought of torture. Elders prevailed and decided to have some fun and give him what he wanted. Sort of. They traded him the swamp land that belonged to our cousins, the Bayogoulas. That's right. Swamp land. (109)

In exchange, the Frenchman gave the Choctaw tools, pots, rifles—and, yes, also the beads. They shared everything with their cousins.

> When my relatives told the Bayogoulas what happened, they went four paws up, laughing, because the land that had been traded was a huge flood plain. Six months out of every year it was knee deep in water, snakes and alligators. Nowhere were there more mosquitoes than on that piece of land. (109–10)

Quite some time passed.

> Then one afternoon a group of Choctaws were tramping through the area and stumbled across *Filan-chi* and his soldiers camping, now get this, in two feet of water. . . . My grandfather called to him from higher ground. . . . "*Filan-chi* what are you doing

down there?" . . . He was flailing his arms like a lunatic and babbling on and on. The gist of his harangue . . . was that the land was his. The end. (110)

Bienville went on to say that

> the bayous had overflowed so furiously that he and his men had been four months in waist high water. My grandfather had to turn away to keep from laughing himself silly. . . . my relatives left him there, soaked to the skin, standing in the middle of "*New France*." And it wasn't until much later that we realized the joke was on us. (110)

Bienville founded New Orleans in 1718. When he saw the natural levees—the result of the sedimentation of the meandering river—in the crescent bend of the Mississippi, he saw protection from river floods, from hurricanes, and from tidal surges. And he saw unlimited promise for trade, being "near the mouth of an enormous system of navigable waterways," with the Gulf of Mexico as "a gateway to the ports of the Americas, Caribbean, and Europe" (Kelman 2003, 6).

In *A River and Its City: The Nature of Landscape in New Orleans*, Ari Kelman calls the Mississippi delta "otherwordly" (2003, 4).

> It is a place of seemingly endless interconnected marshes, swamps, and bayous, with little solid land anywhere in sight. Cattails, irises, mangroves, and a wide variety of grasses thrive in the delta's soggy environment. Muskrats, otters, minks, raccoons, and of course alligators all inhabit this watery world, while crawfish . . . burrow in a constantly replenished supply of muck. Much of the delta remains a wetland wilderness—a great place for fish . . . but a forbidding location for a city. Yet that is exactly what people have done there.

The white settlers, that is. The Indigenous peoples, including groups of Choctaws, who lived in the delta for millennia, did not establish static settlements; they accommodated the dynamic water-land configurations, living and moving with constantly changing constellations. As LeAnne Howe narrates:

> Among Indian tribes in the southeast, there was a continual rhythm of exchange. They gave, we gave. We gave, they gave. That's how things had been done for about 2000 years until *Filan-chi* showed up. I'm not kidding; no one had ever wanted land, forever. This was an anomaly. This changed all rules of government-to-government cooperation. We had no idea how to proceed. (Howe 1994, 109)

The Choctaw had no idea how to proceed in the modernization and colonization of their lands into private property. The settlers had no idea how to proceed in the watery, muddy bayou lands.

One year after its founding, the new settlement experienced its first flood: for six months it was under half a foot of water. The French didn't leave, but dug in. They dug drainage canals in the mud and raised artificial levees on top of the natural ones. It was hard work. Bienville had taken care of that part: he had brought the first African slaves to New Orleans. They did most of the backbreaking labor: "By the 1730s, slave-built levees stretched along both banks of the Mississippi for a distance of nearly fifty miles" (Kolbert 2021, 35–36).

The settlement of the city entailed a new relation: no longer a relational approach of accommodating the fluid water-land dynamics, but one of controlling the water into a river that needed to stay within human-made "banks." The river didn't conform. A game of cat and mouse ensued: miles of higher, broader, and longer levees were erected—and breached. By the twentieth century, the U.S. Army Corps of Engineers was in charge of extensive levee systems, navigation locks, concrete revetments, and engineering feats such as the Old River Control Auxiliary Structure, even more near "otherworldly" than the Mississippi delta itself. "The Corps was not intending to accommodate nature. Its engineers were intending to control it in space and arrest it in time," writes John McPhee in *The Control of Nature* (1989, 10). The Auxiliary Structure was built to keep the Mississippi from jumping course into the Atchafalaya and destroying New Orleans, Baton Rouge, and all the industries built around the river. As McPhee writes, for the Mississippi, such a change of channel was completely natural: it had "jumped here and there within an arc about two hundred miles wide" (5). Major shifts like this happened roughly once a millennium. McPhee doesn't use the term, but Elizabeth Kolbert mentions that these dramatic leaps are called *avulsions*. "Because the Mississippi is always dropping sediments, it's always on the move. As sediment builds up, it impedes the flow, and so the river goes in search of faster routes to the sea" (2021, 33). The Mississippi has avulsed six times within the last seven thousand years (33). However, since the last natural shift, McPhee explains, "Europeans had settled beside the river, a nation had developed, and the nation could not afford nature" (1989, 6). Nature "had become an enemy of the state" (7). At the end of his epic saga of this ongoing battle, McPhee asked a district geologist "if he thought it inevitable that the Mississippi would succeed in swinging its channel west . . . 'Personally, I think it might. Yes. That's not the Corps' position though. We'll try to keep it where it is, for economic reasons'" (92).

One of the consequences of keeping the Mississippi where it is, to keep it from "flooding," jumping course, and spreading its waters and its sediments in dynamic ways, is that Louisiana is one of the fastest-shrinking areas in the world. "Since the 1930s, Louisiana has shrunk by more than two thousand square miles. . . . Every hour and a half, Louisiana sheds another football field's worth of land" (Kolbert 2021, 32). Kolbert adds: "And so a new round of public-works is under way. If control is the problem, then by the logic of the Anthropocene, still more control must be the solution" (32).

Hurricane Katrina made landfall in southeast Louisiana at the end of August 2005. Klaus Jacob, a geophysicist specializing in disaster risk management and resilience to climate change, wrote a radical opinion piece in the *Washington Post* of September 6, 2005, "Time for a Tough Question: Why Rebuild?" According to Jacob, the effect of Hurricane Katrina "is not a natural disaster. It is a social, political, human and . . . engineering disaster" (2005). Moreover, Jacob had warned New York City for years that a hurricane like Sandy might occur. Together with other scientists, he published a report in 2011 carefully describing the threat and the ways the city and the region could protect themselves from storm surges. A year later, Sandy happened.

Jacob deems it inevitable that New Orleans will ultimately fail, and he is not alone: "Government officials and academic experts have said for years that in about 100 years, New Orleans may no longer exist. Period" (Jacob 2005). He suggests a radical revision of the modern approach. Instead of the defensive strategy of the Army Corps of Engineers to protect New Orleans from the water, he advocates a living *with* the water. He envisions a "floating city," or "a city of boathouses, to allow floods to fill in the 'bowl' with fresh sediment" and restoration of wetlands, mangroves, and other buffer zones. Jacob's advice follows basically the way the Choctaws had been accommodating the complexities of the water-land dynamic, living with the changing relations. Let the waters rise and fall. Move with them. Accommodate them. Invite instead of fight. Jacob concludes his opinion piece: "It is time to constructively deconstruct, not destructively reconstruct" (Jacob 2005).

The "joke" was not on the Choctaws, after all.

MENTALITY SHIFT

Constructively deconstructing the model of a rebuilding of New Orleans, as Klaus Jacob suggests, requires a different mindset from the modern mode of controlling and fighting water. It demands a mentality shift toward a relational mode of water as lived and practiced by Indigenous peoples such as the Choctaws. In this last section, I explore what such a shift entails.

I begin with Maurice Merleau-Ponty's understanding of the nature of thinking. In his June 1, 1960, Working Note in *The Visible and the Invisible*, he writes that the fundamental problem of philosophy is sedimentation and reactivation:

> In fact, it is a question of grasping the *nexus*—neither 'historical' nor 'geographic' of history and transcendental geology, this very time that is space, this very space that is time, which I will have rediscovered by my analysis of the visible and the flesh, the simultaneous *Urstiftung* of time and space which makes there be a historical landscape and a quasi-geographical inscription of history. Fundamental problem: the sedimentation and the reactivation. (Merleau-Ponty 1968, 259)

He notes that he has *re*-discovered that time and space are co-originary—that there is a simultaneous fundamental initiation, or, origination (*Urstiftung*) of time and space. Any temporal event has a spatial sedimentation; any sedimentation is partaking in time, changing, moving. *The* fundamental problem is the dynamic between the two: how does stability form and how does renewal or innovation emerge from this stability (Klaver 2016b, 117)?

In the preface to the *Phenomenology of Perception* (1962), Merleau-Ponty emphasizes the importance of renewal by foregrounding the beginning nature of philosophy, the inchoative atmosphere of phenomenology. Invoking Husserl, he states: the "philosopher . . . is a perpetual beginner. . . . It means also that philosophy . . . is an ever-renewed experiment in making its own beginning" (xiv). He concludes: "True philosophy consists in relearning to look at the world" (xx–xxi).

The "relearning to look at the world" is another way of stating that the fundamental problem of thinking is sedimentation and reactivation. Sedimentation and reactivation constitute *meandering*, an invocation of the *re-*, the again. Klaus Jacob's appeal to rethink our way of thinking about the way we design our cities is an example of a reactivation of a sedimentation. Another example is Einstein's saying that we cannot solve our problems with the same thinking we used when we created them. Or, consider Audre Lorde's assertion: "the master's tools will never dismantle the master's house" (Lorde 1984).

The "relearning to look at the world"—no longer by fighting the water with hard barriers, but by working and living with the water—opens up new modes of so-called soft approaches to landscape design projects that will make places such as New York City more resilient against storm surges. No longer fighting the water with hard barriers, soft approaches focus on working and living with the water. The nonprofit Billion Oyster Project works with people across the five boroughs to rebuild the New York Harbor oyster population, which was wiped out in the twentieth century. Restored oyster reefs will function as a buffer to protect the city from wave impact from major storms, to reduce flooding, and to prevent erosion. The project plans to have one billion oysters in New York Harbor by the year 2035.

Similarly, SCAPE Landscape Architecture, with design studios in both New York and New Orleans, designed "Living Breakwaters," partially submerged nearshore rubble mounds that provide habitat for fish and reefs for oysters in the Lower New York Harbor. This shallow estuary once supported commercial fisheries and shell fisheries. The nearest town was historically called "The Town the Oyster Built." The "Living Breakwaters" create a living and dynamic structure that absorbs wave action and prevents further erosion.

These kinds of soft approaches are examples of a larger pulse in innovation in landscape architecture where "the point and counterpoint between positivism and post-positivism, and between art and science, give life to the nature of complexity of adaptation in the built environment" (Keenan 2017, 7).

MEANDER *MÊTIS* RIVERSPHERE

In the following, I relate the nature of complexity to the processes of sedimentation and reactivation in the movement of meandering. I present an approach I have developed in various other works (Klaver 2016; 2017; 2018). This approach embodies a mentality shift: from a modern mentality toward a relational mentality. It entails a different way of thinking based in a meandering mode, a *mētis* mode, a spherical mode. It is not invoked by a crisis, a disaster, but could be a precondition to preventing further disasters insofar as it is no longer an Anthropocene-based mentality of control. Relationality, intrinsic in water, guides the way.

Meandering in its material movement is predicated upon the dynamic relation between water and land/sediment in an ongoing process of sedimentation and reactivation. As we saw above, the Mississippi River is permanently in interaction with the land and waters around it. Meandering conveys the nature of the nonlinear—symbolically and metaphorically. It allows for ambiguity, uncertainty, and hybridity, for that which cannot easily be measured or replicated. Its activity of sedimentation and reactivation is based in the unpredictable workings of the material realm not ruled by structures of scheduled time. Avulsions and storm surges might form occasional extremes, but meandering on every scale is basically unpredictable: it is messier than the straight line. It entails a rethinking of progress through complexity instead of through a modern controlling of nature.

Linearity has been the privileged paradigm of progress and its leading model of efficiency; its concomitant mindset has been goal-oriented or teleological. Bienville's goal was creating a European-style city. Faced with a first inundation, he made water the enemy—the beginning of a long history of fighting the water, rather than inviting it with a changed conception of how to live with the water. Such a change would have been a form of meandering. Convoluted and seemingly undirected, meandering is seen in the modern paradigm not just as the opposite of efficiency, but as being in the way of efficiency. Revaluing meandering has a train of effects on a variety of concepts and practices. It engenders a different way of thinking about efficiency: it might be more efficient in the long term to take more time and explore possibilities, just as a river does when it meanders through a basin. It is a slower process than water running through a concrete channel; it takes more factors into consideration. Making New York's coast more resilient through the "Living Breakwaters" project, for example, entails working with local people, scientists, politicians, biologists, and landscape architects.

Meandering invokes, elucidates, and hints at a different imagination, another mindset, a new epistemological and ontological model, and a cultural and political framework that diversifies what counts as progress and efficiency, as expertise, knowledge, and politics. It bespeaks the social-political necessity of taking time to explore terrains, to elucidate attributes, relations, problems, and solutions, as a gateway to new constructs of imagination.

The movement of meandering is predicated upon an ongoing beginning and reveals how beginning works. Beginning does not take place in a vacuum, is not a *creatio ex nihilo*, but is always building on past experience, or on a break with this experience, as in the case of avulsion, an intensification of reactivation.

Meandering relies on the complex interaction of many material vectors and factors. Its workings are analogous to how complex practices such as knowledge, power, politics, ethics, and aesthetics operate in the everyday: lateral traversing, picking up material and depositing, re-activating in the process. Meandering stands for an ethics of relationality, intertwinement, and entanglement, for a politics of practical engagement, enabling deliberation for experiment, tinkering, and "thinkering," emergent and transient. Meandering brings the social, political, technological, and natural together in an ongoing dynamic. Meandering does not elicit a straight line but a sinuous back and forth, symbolized linguistically by the prefix *re-*, the notion of the again and again, the exploration through wandering, the essay in Michel de Montaigne's original sense of trial and attempt.

Meandering seems to be a slower process than the straight line of progress; yet this is only the case for the simply defined objective, for the short view of time. Meandering proceeds by covering more ground, percolating into deeper depths, listening to the murmurs of more voices, being what it is when and where it is observed. Meandering makes room for what cannot easily be measured, what does not lend itself to be measured, for the slow and the unexpected, and for the workings of the material realm beyond by the structures of scheduled time. Meandering is messy, unpredictable. It echoes Édouard Glissant's indeterminacy as he describes in his *Poetics of Relation*: "The science of Chaos renounces linearity's potent grip and, in this expanse/extension, conceives of indeterminacy as a fact that can be analyzed and accident as measurable" (Glissant 1997, 137).

From early modernity onward, natural meanders in rivers "had to be" engineered away to facilitate major modern projects, such as commercial river transportation and city developments, as exemplified in the endeavors to control and stratify the Mississippi. In the modern mindset, meandering acquired a negative connotation, synonymous with aimless wandering and rambling through a long-winded argument. In the later part of the twentieth century, a reevaluation of meandering emerged: new understandings of chaos and complexity have become foundational in many fields and significant in the cultural imagination (Klaver 2016b). Meandering as a metaphor for a different sort of thinking is founded in and summarizes the nondeterministic models used in many fields of science that were once the hallmark of linear, positivist thinking.

This different mode of thought, which emerges in the revaluing of nondeterministic models, resonates with the ancient Greek *mētis*, a practical intelligence that became overshadowed, backgrounded, and pushed aside by the dominant Greek epistemic of privileging logic. In *Cunning Intelligence in Greek Culture and Society*, Detienne and Vernant (1978) show that *mētis* appears in implicit ways, at "the interplay of social and intellectual customs." It escapes simple definition—it "always appears more or less below the surface, immersed as it were in practical operations . . . applied to situations which are transient, shifting, . . . which do not lend themselves to precise measurement, exact calculation or rigorous logic" (3–4). Detienne and Vernant emphasize that *mētis* materializes as multiple, many-colored, and shifting because:

its field of application is the world of movement, of multiplicity and of ambiguity. It bears on fluid situations which are constantly changing and which at every moment combine contrary features and forces that are opposed to each other. (20)

James Scott, in *Seeing like a State* (2008), demonstrates the significance of *mētis* for academic fields, such as the social sciences, geography, and architecture. He invokes the term *mētis* "to conceptualize the nature of practical knowledge and to contrast it with more formal, deductive, epistemic knowledge" (6). Scott asserts:

There may be some rules of thumb, but there can be no blueprints or battle plans drawn up in advance; . . . such goals can only be approached by a stochastic process of successive approximations, trial and error, experiment, and learning through experience. The kind of knowledge required in such endeavors is not deductive knowledge from first principles but rather what Greeks of the classical period called *mētis*. . . . [It is] the kind of knowledge that can be acquired only by long practice at similar but rarely identical tasks, which requires constant adaptation to changing circumstances. (177–78)

Mētis is the epistemological equivalent of meandering; both are predicated upon a process of relating and adjusting to circumstances. Its ethical equivalent is an ethics of relationality instead of a rule-based ethics.

Mētis had already become backgrounded in the Greek intellectual and philosophical world. In modernity it all but disappeared as a legitimate mode of knowledge, replaced by the expertise of engineers, scientists, and designers whose models of nomothetic-deductive logic appeared very efficient and successful, but often came with long-term devastating consequences. As Scott explicates: "The utilitarian commercial and fiscal logic that led to geometric, mono-cropped, same-age forests also led to severe ecological damage" (309). A similar logic of apparent initial efficiency but long-term *in*efficiency and harm emerges in the creation of modern water, such as in dam building and draining of swamps.

Detienne and Vernant show how a *mētis* mode of thinking is closer to Chinese and Indian modes of thought than to Greek philosophy, which is characterized by a dichotomy between being and becoming as follows:

On the one hand there is the sphere of being, of the one, the unchanging, of the limited, of true and definite knowledge; on the other, the sphere of becoming, of the multiple, the unstable and the unlimited. . . . Within this framework of thought there can be no place for *mētis*. *Mētis* is characterised precisely by the way it operates by continuously oscillating between two opposite poles. (5)

Mētis is beyond dualistic thought, like meandering with its movement of sedimentation and reactivation. Through a *mētis* lens, water emerges in its polydimensional, nondeterministic, and dynamic character. This includes a multispherical dimension. Rivers are more than lines on a map such as Viele's, more than their basins, watersheds, or drainage areas. They influence the geology, the air, and the soil around them, life around them, cultures around them. They create their own hydrosphere, biosphere, and atmosphere. They form intricate networks of relations, conditions of possibilities.

I specify the concept of riverine atmosphere as *riversphere*, to examine rivers as places of multiscalar and multivector connectivity and complexity. My sense of riversphere resonates with Gernot Böhme's (1993) concept of atmospheres:

> Atmospheres are indeterminate above all as regards their ontological status. We are not sure whether we should attribute them to the objects or environments from which they proceed or to the subjects who experience them. We are also unsure where they are. They seem to fill the space with a certain tone of feeling like a haze." (Böhme 1993, 114)

Riversphere is a thick, profoundly relational concept. It negotiates and blurs separate spheres—such as hydrosphere, geosphere, atmosphere—and adds social, political, cultural, aesthetic, and affective dimensions to our water conceptualizations and praxes. It enriches the conceptualization of rivers in the cultural imagination, intertwining hydrological, biological, ecological knowledge and experience with lived experience, social-cultural and political activities, storytelling, and more. LeAnne Howe's piece on the founding of New Orleans by *Filan-chi* (Bienville) gives a rich example of the relationality of multiple spheres embedded in narrative. This spheric relationality can also pertain to hybrid waters, such as infrastructural waters. Nikhil Anand (2017) develops a notion of hydraulic citizenship predicated upon the deep intertwinement, the entanglement, of the dynamic of infrastructural water in pipes and pumps, with citizens, technicians, politicians, plumbers: a complex vibrant and relational mix of stories, theories, facts, and experiences. In "Accidental Wild," I describe how the hybrid water of a flood control detention pond becomes a place for multicultural and multispecies encounters (Klaver 2015). The precondition for this relationality to happen is to not overcontrol the area, but to leave it relatively wild, indeterminate. At the same time, such a hybrid model assures that the detention pond has enough room to rise and fall in the case of intense rain events and the rise of the creek, and to prevent flooding in town, which is designed, as are most modern towns, with concrete channels and impervious surfaces of streets and parking lots.

Jamie Linton, who coined the term *modern water*, shows how in the reduction of water to modern water the hydrosphere has become a strictly separated domain from the socio-sphere: "the hydrological cycle conditions an understanding that keeps water and people in separate, externally related spheres" (106). Within a meander, *mētis*, and riversphere approach, geometrical and homogenizing models of water give way to models of complexity and indeterminacy (Klaver 2017), thereby giving room to multiple materialities and relationalities.

Based upon their work in the Lower Mississippi River Valley in the 1990s, landscape architects Anuradha Mathur and Dilip da Cunha came to see the river as an invention of colonizing practices in which land and water have become strictly separated. Da Cunha convincingly elaborates this perception in his book *The Invention of Rivers: Alexander's Eye and Ganga's Descent*. He contrasts the line of the river Ganges with the ubiquity of the rain-driven wetness of the goddess Ganga's descent from heaven. It contrasts a thinking in terms of unity of rivers with that of the indeterminacy of rain. He invokes a new imagination anchored in rain, in Ganga's descent, "one that drives the design of new infrastructure and an alternate edifice of myths, facts, ideas, practices and frameworks of critique" (2019, 293). Da Cunha's analysis entails a radical relearning of looking at the world.

> Ganga does not flow as the Ganges does, in a course to the sea; she is rather held in soils, aquifers, glaciers, living things, snowfields, agricultural fields, tanks, terraces, wells, cisterns, even the air, all for a multiplicity of durations that range from minutes and days to centuries and eons. She soaks, saturates, and fills before overflowing her way by a multiplicity of routes. . . . The only anchor she offers people is the time of her descent. It is celebrated each year at the coming of the monsoon. (40)

Da Cunha contrasts Ganga's descent with the invention of the river Ganges, created by Alexander's eye, that is, the eye of the conqueror Alexander the Great of Macedon. In 334 BCE he set out eastwards, not only with his army of soldiers but with an expedition of scholars, scientists, zoologists, surveyors, artists, and historians, collecting "new cartographic data" (25–27).

> His campaign gathered information for science, described places, and affirmed ideas. More seriously, however, it called out a ground—an earth's surface constituted of land and water to begin with—that . . . was 'unknown even to the Indi.' It was perhaps Alexander's most lasting legacy . . . It involved articulating things with a line that could be drawn on a map, more conveniently perhaps than on the ground. (27)

That line was the river. Alexander did not reach the Ganges, but gave the river its name, drew it on a map, brought it into existence (29). Still today, two millennia later, the lines he drew, the rivers he created, are "an essential feature not just in maps of India but on the ground in riverbanks, riverfront projects, regulations, and flood control schemes" (30).

With the dominant creation of water in the shape of a line—a river—a worldview of dualistic thinking developed, including the dualism of land versus water. Only when such a line is drawn do floods appear. Floods don't exist beyond the line. Da Cunha radicalizes Klaus Jacob's stance that Hurricane Katrina is not a natural disaster but a man-made social, political, and engineering disaster. For da Cunha there are no natural disasters, only design disasters. Jamie Linton argued a similar position: there are not crises of water per se, but only crises of *modern* water.

Da Cunha is convincing in his presentation of the river as colonial invention. However, one can conceive of rivers as contingent emerging wholes, which are not necessarily to be seen as unitary. Geographers Philip Steinberg and Kimberley Peters (2015) present this possibility with their notion of "wet ontology," which has ontological ramifications similar to those I develop in the meander-mētis-riversphere approach. Moving away from a "linear and lateral narrative" in geology toward "complexity-based understandings of chaos-inspired" accounts, they come to a Gilles Deleuze–informed "'assemblage' approach that presupposes a world of immanence and becoming. . . . Key to an assemblage is that the parts of which it is composed are heterogeneous and independent, and it is from the *relations* between the parts that the temporary, contingent whole emerges" (225, emphasis in original). They refer to Stephanie Lavau's work on sustainable water management in Australia as embodying their "'wet ontological' perspective, in which flow is, on the one hand, a singular force but, on the other hand, composed of multiple chaotic processes. For Lavau, water, in both its singular and multiple existences, incorporates and confounds human intervention" (257). Management strategies that allow for a coexistence of multiple "ontologies of thought" reflect "water's persistence as a vibrant matter that has agency in its 'unruliness, variability, mobility, and fluidity'" (257). In these approaches, too, relationality is the key to escaping the homogenizing and unifying force toward modern water. As Lavau states, cited by Steinberg and Peters: "Embracing relational materiality leads us to ontological multiplicity, to attending to the different realities that are produced in particular socio-material orderings" (257).

RADICAL RELATIONALITY

There are convergent resonances in the many positions presented here. A common thread is formed by relationality. All of them acknowledge an agency of water in and through relationality.

Water offers counterontologies, counterepistemologies, counterethics, and counteraesthetics, radically rooted in relationality, including the intrinsic relation between water and land, decentering the human and the notion of the individualized subject. Water is not radical because it is root-like. It is radical because it does not allow itself to be reduced to one root, to simplicity. It embodies a relational being, knowing, thinking, judging, designing, and living *with*. These ways with water accommodate and follow the multiple ways water is, travels, acts, relates and cognates. This takes us out of a language of containment, out of rigid ways of categorizing, into regenerating flexibility, places of messiness in orders. It entails listening for the granular grammar of water, becoming lost in translations of its countless dialects, its rain and waterfalls, its almost inaudible hush of being pumped through mazes of capillaries in plant, animal, human, and infrastructural bodies, under the ground, through mountain veins, gurgling, writing its hieroglyphs in the mud, in sandstone.

Radical water teaches us to take water seriously, not because we are afraid of it, not because it has been made the Other, the enemy that is out to get us, but rather to take water seriously because we know that it is relational, knowing that water guides the way and thereby becomes radical water. In its fluid gentleness, its violent persistence, it teaches us relationality, it teaches us to change: to live *with* it, instead of controlling it. This entails changing and innovating not just technologies and designs, but ways of thinking, being, acting, engaging. Water teaches relationality in our decisions, interventions, conscious changes in policy, economics, culture. Radical water teaches us to take water seriously, that is, relationally; we can't just "take it" anymore.

REFERENCES

Ahmed, Sara. 2017. *Living a Feminist Life*. Durham, NC: Duke University Press.

Anand, Nikhil. 2017. *Hydraulic City: Water and the Infrastructures of Citizenship in Mumbai*. Durham, NC: Duke University Press.

Associated Press. 2020. "Mayor: New York City Seal Should Be Re-Examined." July 28, 2020. www.fox5ny.com/news/mayor-new-york-city-seal-should-be-re-examined.

Böhme, Gernot. 1993. "Atmosphere as the Fundamental Concept of a New Aesthetics. *Thesis Eleven* 36: 113–26.

Culliton, Kathleen. 2020. "Mayor Would Consider Changing NYC Seal Depicting Native American." NY1 Spectrum News, July 27, 2020. www.ny1.com/nyc/all-boroughs/news/2020/07/27/mayor-considers-changing-nyc-seal-.

Da Cunha, Dilip. 2019. *The Invention of Rivers: Alexander's Eye and Ganga's Descent*. Philadelphia: University of Pennsylvania Press.

Detienne, Marcel, and Jean-Pierre Vernant. 1978. *Cunning Intelligence in Greek Culture and Society*. Translated by Janet Lloyd. Chicago: University of Chicago Press.

Eckstrom, Jim. 2020. "Seneca President: NYC Seal 'Example of Outdated and Racist Imagery.'" *Olean Times Herald*, July 30, 2020. www.oleantimesherald.com/news/seneca-president-nyc-seal-example-of-outdated-and-racist-imagery/article_ed68c6be-680f-5815-8024-d202d91c4278.html.

Estes, Nick. 2019. *Our History Is the Future: Standing Rock versus the Dakota Access Pipeline, and the Long Tradition of Indigenous Resistance*. London: Verso.

Glissant, Édouard. 1997. *Poetics of Relation*. Translated by Betsy Wing. Ann Arbor: University of Michigan Press.

Goodell, Jeff. 2017. *The Water Will Come: Rising Seas, Sinking Cities, and the Remaking of the Civilized World*. New York: Little, Brown.

Howe, LeAnne. 1994. "The Chaos of Angels." *Callaloo* 17 (1): 108–14.

Jacob, Klaus. 2005. "Time for a Tough Question: Why Rebuild?" *Washington Post*, September 6, 2005. www.washingtonpost.com/wp-dyn/content/article/2005/09/05/AR2005090501034.html.

Kaika, Maria. 2005. *City of Flows: Modernity, Nature, and the City*. London: Routledge.

Keenan, Jesse M. 2017. Introduction to *Blue Dunes: Climate Change by Design*, edited by Jesse M. Keenan and Claire Weisz, 5–23. New York: Columbia Books on Architecture and the City.

Kelman, Ari. 2003. *A River and Its City: The Nature of Landscape in New Orleans*. Berkeley: University of California Press.

Klaver, Irene J. 2015. "Accidental Wildness on a Detention Pond." *Antennae* 33 (Autumn): 45–58. www.antennae.org.uk.

———. 2016a. "Water, Mud, and Sand: Dutch Re-scaping the Land." In *Hypernatural Landscapes in the Anthropocene*, edited by Sabine Flach and Gary Sherman, 101–22. Oxford: Peter Lang.

———. 2016b. "Re-Rivering Environmental Imagination: Meander Movement and Merleau-Ponty." In *Nature and Experience: Phenomenological Approaches to the Environment*, edited by Bryan E. Bannon, 113–27. London: Rowman and Littlefield.

———. 2017. "Indeterminacy in Place: Rivers as Bridge and Meandering as Metaphor." In *Phenomenology and Place*, edited by Janet Donohoe, 209–25. London: Rowman and Littlefield.

———. 2018. "Re-Claiming Rivers from Homogenization: Meandering and Riverspheres." In *From Biocultural Homogenization to Biocultural Conservation*, edited by Ricardo Rozzi et al., 49–69. Dordrecht, The Netherlands: Springer Press.

Klaver, Irene J., and J. Aaron Frith. 2014. "A History of Los Angeles's Water Supply: Towards Reimagining the Los Angeles River." In *A History of Water*, series 3, vol. 1, *From Jericho to Cities in the Seas: A History of Urbanization and Water Systems*, edited by Terje Tvedt and Terje Oestigaard, 520–49. London: I.B. Tauris.

Kolbert, Elizabeth. 2021. *Under a White Sky: The Nature of the Future*. New York: Crown.

Kurutz, Steven. 2006. "When There Was Water, Water Everywhere." *New York Times*, June 11, 2006. www.nytimes.com/2006/06/11/nyregion/thecity/11viel.html.

Linton, Jamie. 2010. *What Is Water? The History of a Modern Abstraction*. Vancouver: UBC Press.

Lorde, Audre. 1984. "The Master's Tools Will Never Dismantle the Master's House." In *Sister Outsider*, edited by Audre Lorde, 110–13. New York: Crossing Press.

Mathur, Anuradha, and Dilip da Cunha, eds. 2014. *Design in the Terrain of Water*. San Francisco: Applied Research + Design.

McPhee, John. 1989. *The Control of Nature*. New York: Farrar, Straus and Giroux.

Merleau-Ponty, Maurice. 1962. *The Phenomenology of Perception*. Translated by Colin Smith. London: Routledge and Kegan Paul.

———. 1968. *The Visible and the Invisible*. Edited by Claude Lefort, translated by Alphonso Lingis. Evanston: Northwestern University Press.

Mulholland, Catherine. 2000. *William Mulholland and the Rise of Los Angeles*. Berkeley: University of California Press.

New York City Seal. 2015. Colored rendering by K. Lefebvre of Paul Manship's 1915 authorized sculpted version of the seal. https://commons.wikimedia.org/wiki/File:Seal_of _New_York_City.svg, accessed March 7, 2021.

NYC Green Book Highlights. N.d. City Seal. www1.nyc.gov/site/dcas/about/green-book -city-seal-and-flag.page, accessed January 8, 2020.

Sartin, Jeffrey S. 1993. "Infectious Diseases during the Civil War: The Triumph of the 'Third Army.'" *Clinical Infectious Diseases* 16: 580–84.

Schuman, Melanie. 2020. "New York City Mayor Warns Residents to Take the Storm 'Very Seriously.'" CNN, August 4, 2020. https://edition.cnn.com/us/live-news/isaias-storm -08-04-2020/h_59c479e6e16bbf65011fe2ad7f26471f.

Scott, James. C. 1998. *Seeing like a State: How Certain Schemes to Improve the Human Condition Have Failed*. New Haven: Yale University Press.

Segovia, Jose Francisco. 2010. "Vielé, Egbert Ludovicus (1825–1902)." In *Handbook of Texas Online*, Texas State Historical Association, uploaded on June 15, 2010. www.tshaonline .org/handbook/online/articles/fviya.

Simpson, Leanne Betasamosake. 2017. *As We Have Always Done: Indigenous Freedom through Radical Resistance*. Minneapolis: University of Minnesota Press.

Steinberg, Philip, and Kimberley Peters. 2015. "Wet Ontologies, Fluid Spaces: Giving Depth to Volume through Oceanic Thinking." *Environment and Planning D: Society and Space*, no. 33: 247–64.

Vielé, Egbert L. 1865. *Sanitary & Topographical Map of the City and Island of New York Prepared for the Council of Hygiene and Public Health of the Citizens Association*. Under the direction of Egbert L. Viele, Topographical Engineer. New York: Ferd. Mayer & Co. Lithographers. Retrieved from David Rumsey Historical Map Collection in the David Rumsey Map Center of Stanford University. www.davidrumsey.com/rumsey/download .pl?image=/D0018/3723000.sid target=_blank.

Vielé, Egbert L. and Ferd. Mayer & Co. 1865. *Topographical Map of the City of New York: Showing Original Water Courses and Made Land*. New York: Ferd. Mayer. Map. Retrieved from the Library of Congress, www.loc.gov/item/2006629795/.

Yazzie, Melanie K., and Cutcha Risling Baldy. 2018. "Introduction: Indigenous Peoples and the Politics of Water." *Decolonization: Indigeneity, Education & Society* 7 (1): 1–18.

NOTES

1. During summer 2020, after many racially-charged events, various cities and organizations in the nation began questioning racist imagery in official monuments, documents, and naming practices. New York City mayor Bill de Blasio expressed willingness to reconsider the imagery on the seal of New York City. Executive director of the Lenape Center Joe Baker and President of the Seneca Nation Rickey Armstrong Sr. both welcomed de Blasio's intent. Baker: "The seal ignores the history of violence and destruction inflicted on Indigenous people by settlers" and "presents a caricature and negative representation of Native culture." Armstrong: "It is our hope that . . . the conversations taking place around these important issues in New York and across the nation will lead to greater respect, fairness and justice for Native people and our rich culture" (Culliton 2020; Eckstrom 2020; Associated Press 2020).

PART II

Fluid Identities

WATER, IN ITS MANY FORMS—FLUID AND FIXED, culture and nature, substance and abstraction—is entangled in a wealth of scholarship exploring meaningful relationships between bodies both human and aquatic. Water is intimately tied to collective identites at different scales, including ethnic, national, and imperial (Blum 2010; Bose 2016; DeLoughrey 2007), and is an especially productive medium for exploring how cultures travel and transform (Dawson 2018; Gilroy 1993; Somerville 2012). Water's own identities are also in flux. It is at once an essential component of the human bodies and cultures that it constitues and mirrors, while also powerfully exceeding them (Linton 2010; Helmreich 2011). Oceans, for example, can be simultaneously construed as social spaces (Cusak 2014; Steinberg 2001) and alien domains (Helmreich 2009). The very boundaries between land and water can be tools of settler colonial power, as the elemental categories of Western ontologies are enlisted in extractive regimes (Sammler 2020). Or, in Stacy Alaimo's (2012) oceanic conceptualization of transcorporeality, material boundaries are permeable and to be understood as exchanges connecting bodies of humans, animals, and beyond, in relationships that cannot be disassembled (476).

The three chapters in this section deal with how the presence or absence of water is powerfully entangled with ethnic, national, and imperial cultural identities, highlighting the interconnectedness between human and material worlds. Chapter 4 brings the agency of water into conversations of identity. Ignacio López-Calvo and Hugo Alberto López Chavolla look, from the perspective of new materialism, at Peruvian and Colombian Indigenous worldviews in relation to water in the novels of two non-Indigenous authors. Swelling montane rivers have agency to serve as a role model for emancipation to subdued Quechua communities in the Andes; music and dance inspired by water remind the novel's protagonist about the deep Indigenous roots of his nation; rivers even converse with characters in troublesome ethnic identitarian crises and lead the way to survival through dry seasons. Explorations of literary representations take on material and political significance in the existential survival of ethnic groups such as the Wayuu, in La Guajira Peninsula of Colombia, where extractivism and the neoliberal policies of international mining corporations, in collusion with local authorities, have been damaging the health, morale, and self-esteem of Indigenous groups for decades. The lack of drinking water after the corporation takes over the river drives young Wayuu people to move to the cities or to work for the very corporation that is destroying their ancestral way of life. And, once they leave their ethnic communities, the danger of forgetting their own language and traditions is dramatically increased. Both novels incite the reader toward praxis, toward a decolonial activism against Criollo oppression in the first case, and against international, neoliberal, predatory extractivism in the second. Both novels likewise promote Indigenous ethnic pride, solidarity, and unity as a way to overcome sociopolitical and economic adversity, as well as to preserve, recover, and promote endangered cultural identities.

In chapter 5, Penelope K. Hardy focuses on the identities assigned to the ocean by cartographers whose priorities when attempting to map the global seabed tended to coincide with the commercial and political interests of empire. Hardy explores the intricate connections between ocean travel and global imperialism, particularly through the process of naming sea features as a mark of possession. As she explains, after Germany's defeat in World War I, German scientists, realizing that they had lost all their colonies where fieldwork used to be conducted in a friendly environment, saw in the bottom of the ocean a new area where they could conduct prestigious research experiments as well as an area that could be symbolically claimed by Germany. The author points out: "German oceanographers used their technologically derived knowledge of the ocean bottom to assert their—and their nation's—continued membership in the top tier of science and thus their continued claim to be a Great Power." Decidedly terrestrial practices for naming and claiming land are imposed onto seascapes, comingling their identities in the pursuit of empire.

In the closing chapter of the section, Kale Bantigue Fajardo studies how water is being enlisted in transforming the national imaginary in the Philippines. Focusing on the city of Malalos, considered the birthplace of the nation in 1898, Fajardo shows how its identity since independence has been reoriented away from land-based imaginaries and toward water-based ones. Cultural production, tourism, and urban development are shifting the focus beyond inland colonial architecture to include maritime images of canoes, rivers, estuaries, coasts, and islands. This process presents a more-than-human precolonial and decolonial counterpoint to hegemonic nationalism. Indeed, the ethno-linguistic roots of Tagalog are decidedly aquatic, meaning "from the river." Fajardo celebrates this move toward "aquapelagic" narratives, together with the historical archiving of the area, particularly in light of the possibility that climate change may one day bring flooding to the area.

Together, these three chapters contemplate how water, from mountain streams to oceanic depths, affects human identities and survival. Human relationships with rivers and oceans over the centuries have transformed ways of being in the world in dramatic ways. Water, after all, is a fluid reminder for some populations—who sometimes see water as sacred rather than as a commodity—of their Indigenous roots. For others, the existence of rivers is equated with the survival of entire cultures threatened by drought and neoliberal extractivism. Thus, fighting against altering the course of a stream or river may be seen as a struggle against internal colonialism. Powerful countries see in the oceans and seabeds possibilities for national pride, scientific prestige, and even imperial expansion. And water is, of course, also framed as a tourist attraction to improve local economies while concomitantly being considered a destructive threat under climate change conditions: a city always imagined as a site against colonial domination may begin to be conceived in ecological or touristic terms. Different disciplinary approaches, such as literary studies, history, cartography, ethnic and urban studies, join forces to answer key questions about cultural identities amidst environmental uncertainties. This section, therefore, moves from theoretical approaches to policy change recommendations, including engaged literature and pro-Indigenous activism that attempts to provide a voice for subaltern groups.

REFERENCES

Alaimo, Stacy. 2012. "States of Suspension: Transcorporeality at Sea." *Interdisciplinary Studies in Literature and Environment* 19 (3): 476–93.

Blum, Hester. 2010. "The Prospect of Oceanic Studies." *Proceedings of the Modern Language Association* 125 (3): 670–77.

Bose, Sugata. 2006. *A Hundred Horizons: The Indian Ocean in the Age of Global Empire.* Cambridge, MA: Harvard University Press.

Cusak, Tricia. 2014. *Framing the Ocean, 1700 to the Present: Envisaging the Sea as Social Space.* London: Routledge.

Dawson, Kevin. 2018. *Undercurrents of Power: Aquatic Culture in the African Diaspora.* Philadelphia: University of Pennsylvania Press.

DeLoughrey, Elizabeth. 2007. *Routes and Roots: Navigating Caribbean and Pacific Island Literatures.* Honolulu: University of Hawaii Press.

Gilroy, Paul. 1993. *The Black Atlantic: Modernity and Double Consciousness.* Cambridge, MA: Harvard University Press.

Helmreich, Stefan. 2009. *Alien Ocean: Anthropological Voyages in Microbial Seas.* Berkeley: University of California Press.

———. 2011. "Nature/Culture/Seawater." *American Anthropologist* 113 (1): 132–44.

Linton, Jamie. 2010. *What Is Water? The History of an Abstraction*. Vancouver: University of British Columbia Press.

Sammler, Katherine. 2020. "Kauri and the Whale: Oceanic Meaning and Mattering in New Zealand." In *Blue Legalities: The Life and Law of the Sea*, edited by Irus Braverman and Elizabeth R. Johnson, 63–84. Durham, NC: Duke University Press.

Somerville, Alice Te Punga. 2012. *Once Were Pacific: Māori Connections to Oceania*. Minneapolis: University of Minnesota Press.

Steinberg, Philip E. 2001. *Social Construction of the Ocean*. Cambridge: Cambridge University Press.

4

Water, Extractivism, Biopolitics, and Latin American Indigeneity in Arguedas's *Los ríos profundos* and Potdevin's *Palabrero*

Ignacio López-Calvo and Hugo Alberto López Chavolla

Is it not possible to imagine matter quite differently: as perhaps a lively materiality that is self-transformative and already saturated with the agentic capacities and existential significance that are typically located in a separate, ideal, and subjectivist, realm? . . . Is it possible to understand a process of materialization and the nature of its fecundity, to grasp matter's dynamic and sometimes resistant capacities, without relying upon mysticisms derived from animism, religion, or romanticism? (Diane Coole 2010, 92)

This chapter studies the significance of water for Latin American Indigenous communities at two different levels. First, using the theoretical perspective of new materialism (the agency of objects), it concentrates on the symbolic, cultural, magical, and salvational significance of water (of montane rivers in particular) for Quechua communities in Peru, as re-created in Peruvian José María Arguedas's (1911–69) novel *Los ríos profundos* (*Deep Rivers*, [1958] 2004). This worldview is contrasted, from an ecocritical perspective, with the importance of water from a sustainability perspective for the Wayuu Indigenous group in the Guajira Peninsula of Colombia, as represented in the Colombian Philip Potdevin's (1958–) *Palabrero* (2016). This novel re-creates the real-life fact that, since an international mining company began to steal water resources from the Wayuu's ancestral lands three decades ago, suicide rates have grown exponentially among the members of this Indigenous group.

This chapter looks at both novels from the perspective of neomaterialist theory. Diana Coole and Samantha Frost, in the introduction to their 2010 edited volume

New Materialisms: Ontology, Agency, and Politics, propose to acknowledge the agency of things:

> Conceiving matter as possessing its own modes of self-transformation, self -organization, and directedness, and thus no-longer as simply passive or inert, disturbs the conventional sense that agents are exclusively humans who possess the cognitive abilities, intentionality, and freedom to make autonomous decisions and the corollary presumption that humans have the right or ability to master nature. (10)

Neomaterialism, therefore, takes for granted that, along with humans (and independently from humans), things are also agents and can offer resistance through an existing "sociocultural territory composed by relations among the people and earth-beings, *and* demarcated by a modern regional state government" (de la Cadena 2015, 5).[1] In fact, these theorists argue that it is precisely the Western division between the human as agent and the thing as object (or instrument) that has served as a veiled justification for the centuries-long ecological destruction of the planet. In the same vein, Marisol de la Cadena, in her 2015 book *Earth Beings: Ecologies of Practice across Andean Worlds*, asserts:

> Conjuring earth beings up into politics . . . may indicate that nature is not only such, that what we know as nature can be society . . . Representation can make the world legible as one and diverse at the same time by translating nature (out there everywhere) into the perspectives of science (the universal translator) . . . To be able to think 'earth-beings,' the world that underwrites the distinction between nature and humanity requires a translation in which earth-beings become cultural belief. (99)

From this perspective, they explore the potential role of materiality for political constitution. In our opinion, this theory is certainly reminiscent of traditional native worldviews throughout the Americas in which humans are conceived of, without exceptionalisms, as an intricate part of the natural world, rather than as an owner who can use its resources at will. The organic and inorganic, human and nonhuman are, therefore, situated on an equal footing in terms of political agency, against the assumption that only the rational human is capable of it: "While new materialists' conceptualization of materialization is not anthropocentric, it does not even privilege human bodies. There is increasing agreement here that all bodies, including those of animals (and perhaps certain machines, too), evince certain capacities of agency" (Coole 2010, 20). Things evince not only agency but also political agency because, as Frost and Coole explain, they can affect the structure of political life. From this perspective, we will now analyze the potential agency of water as seen by Indigenous worldviews in two Latin American novels.

NEW MATERIALSISM AND *LOS RÍOS PROFUNDOS*

In the novel *Los ríos profundos* (1958), José María Arguedas presents the semiautobiographical story of the protagonist Ernesto and his journey into becoming a young man, after his father, a traveling lawyer, leaves him for boarding school in the town of Abancay, under the direction of Father Linares, a Catholic priest. As Ernesto becomes immersed in the boarding school, a microcosm of Peruvian society, alongside other children and teenagers, he engages in a process of observation, followed by deep reflection, which will shape his conscience as he discovers his place in the world. Although, due to the structure of the novel, it is difficult to fully grasp with precision the origins of Ernesto and his father, Arguedas makes it clear that Ernesto is a Mestizo who spent most of his early childhood traveling alongside his father through the different regions of Peru and in close contact with Quechua communities. In the process, he learned their language, Quechua, and developed a deep respect for Quechua traditions and culture.

In the same manner, and following the Quechua Indigenous worldview, Ernesto develops an awareness of his own interconnectedness with the rest of nature, including rivers, animals, plants, and mountains. These experiences, rooted in Ernesto's early years, allow him to serve as a bridge—or rather as a river—between cultures: the Criollo and the Indigenous ones. In this sense, he becomes "a composition (perhaps a constant translation) in which the languages and practices of [his] worlds constantly overlap and exceed each other" (de la Cadena 2015, 5). At the same time, however, both Quechua people and Criollos marginalize him, as his transculturated nature prevents him from being fully able to insert himself into either culture. Ultimately, however, Ernesto attempts to position himself on the side of the Indigenous cause and, in many aspects, he ends up rejecting the Western heritage that has been violently imposed upon the Quechua communities. This rejection of Western culture grows with his increased awareness of Indigenous oppression, as the novel follows the cadence of a bildungsroman. Here, it is important to note that Ernesto's growth of conscience appears to be linked to the communal indignation of groups, one Mestizo and one Indigenous, that appear in the novel: the *chicheras* and the *colonos*, respectively.[2]

Beginning with the title, water, generally in the form of rivers, is a significant and recurrent element that is presented not solely as part of the scenery but also as an actor with its own agency. In this sense, through the symbolic presence of water (Andean deep rivers), Arguedas evokes the depth—the solid, ancestral roots and matrices of the national identity of Peru—of the Andean culture, as opposed to the superimposed character of a Western and cosmopolitan culture behind the thousand-year-old historical legacy of native Peru.[3] These rivers that shape and configure Peruvian topography then become markers that delineate the cultural contours and historicity of the Andean region. Furthermore, as the story progresses, the presence of rivers in the novel becomes charged with promises of a future in which Quechua people will recover the dignity that has been suppressed by centuries of marginalization at the hands of their oppressors. Arguedas hints at this message, among other times, when he chooses the title of "Yawar mayu" (Blood River) for the tenth, penultimate, and longest chapter of his novel. In it, the native colonos revolt for the first time, demanding that Father Linares hold mass and pray after a disease similar to the Black Death scourges Abancay as well as nearby towns and villages.

Although an earlier revolt led by the chicheras has already taken place in chapter 7, "El motín," not only is it unsuccessful but it also provokes the militarization of Abancay in order to prevent further revolts. Notwithstanding the failure of this first revolt, by insisting on a second revolt taking place in chapter 10 with a Quechua title, and also by implying a connection between nature and human life, Arguedas urges for the recognition of both Indigenous and Western visions of the universe. Thus, the title "Yawar mayu" and the occurrences within the chapter remind us that our existence depends from one moment to the next, not only on socioeconomic structures that produce and reproduce the conditions of our everyday lives, but also on a myriad of microorganisms and diverse higher species, on our own hazily understood bodily and cellular reactions and on pitiless cosmic motions, on the material artifacts and natural stuff that populate our environment (Coole 2010, 1).

Using Ernesto as both bridge and river between Quechua and Criollo cultures, Arguedas "alludes to the fact that social and cultural constructions that are truly solid and authentic must be nourished (as the coastal rivers that drink from montane rivers do) of Andean, Indigenous elements, which represent the fertile and synthesizing miscegenation of Peru."[4] In other words, just as Andean rivers bring the essence of Quechua indigeneity to the rest of the country, so will Ernesto fight for the survival of Indigenous cultures and for the respect they deserve in Peru. As the protagonist learns his place in the world, his confusion about who he is and where he stands is a constant trope that is often associated with the manner in which he interacts with his surroundings, rivers in particular, which once again connote Indigenous cosmovision and reiterate notions of the agency of nature: "I did not know if I loved the bridge or the river more. But both cleared my soul, flooding it with strength and heroic dreams. All weeping images, doubts and bad memories were erased from my mind."[5] Here, the bridge and the river provide Ernesto with peace of mind because both of them represent the unification of two cultures.

On the one hand, the river, whose existence within Peru is undeniably legitimate, can represent the very life of the Andean cultures running through the earth, like blood through our veins; on the other, the bridge, which was forced into existence by human action, was constructed with materials and techniques of Spanish origin, and represents the Spanish cultural heritage imposed on Peru. Even though they might represent opposites in a cultural sense, the fact that Ernesto is at ease while contemplating them and cannot make up his mind about which one he loves the most, suggests an attempt by Arguedas at synthesizing a discourse that exalts the virtues of Indigenous cosmovision, while still being intrinsically connected with Spanish heritage and tradition.

In the same vein, Ernesto's childhood experiences with Quechua communities and his being rejected by some of his schoolmates because of his Indigenous cultural background encourage him to adopt Indigenous worldviews in order to make sense of his own confusing reality. As Ernesto is left in school and his father leaves once again, Arguedas writes: "[Ernesto] would receive the powerful and sad current that hits the children, when they must face alone a world full of monsters and fire and great rivers that sing with the most beautiful music when hitting stones and islands."[6] The music of rivers, then, consoles the aching teenager. The disheartening description of Ernesto's entrance into an unknown world by himself is promptly soothed by the reassurance of the presence of rivers that sing the most beautiful melodies. These rivers, thus, become both hope and life for Ernesto. As his father departs, Ernesto, through a song, enters into a dialogue with nature, where the presence of water is crucial:

> Do not forget, my little one, do not forget!
> White hill, make it come back;
> Mountain water, spring of the pampa that never dies of thirst.
> Falcon, carry it on your wings and make it come back.
> Immense snow, father of snow, do not hurt it on the road.
> Bad wind, do not touch it.
> Storm rain, you do not reach it.
> No, precipice, atrocious precipice, do not surprise him!
> My son, you must return, you must return![7]

Through this prayer-like song, Ernesto engages with nature and asks it to be kind to his traveling father. Simultaneously, the presence of water is significant, as it appears as both a curse and a blessing. For one, water could serve his father as sustenance so that he does not suffer thirst, but at the same time it represents imminent dangers, as it could present itself as snow or storm, which could be deadly. As in this passage, in several others we see Ernesto addressing nature and speaking directly to sacred rivers, immersed in a world of magic and myth. This act, in and of itself, suggests that the Andeanized Ernesto believes in the agency of a dynamic natural world. In neomaterialist terms, it "means returning to the most fundamental questions about the nature of matter and the place of embodied humans within a material world" (Frost and Coole 2010, 3).

Arguedas, therefore, inserts elements of Andean culture (mythical-magical mentality, Quechuanized Spanish, oral tradition, folk music and dance) within Western cultural forms (including among them the Spanish language and the genre of the novel), to a degree of transculturation greater than that achieved by other authors linked to the Indigenista or Neo-Indigenista literary movement.[8] In other words, rather than using fantasy as a conduit to understand the reality of the Andean region, Arguedas takes the fluid reality of Andean language in concert with an Indigenous cosmovision, and then juxtaposes it with the traditions brought by the Spanish in order to highlight the inevitable clash between the two cultures. By the same token, this contact, simmering over five centuries, has led to a mixture of cultures that, for Arguedas, is key to understanding Peru's cultural present and, perhaps more important, to build an inclusive and harmonious cultural future for the nation.

From the perspective of Quechua cosmovision, in *Los ríos profundos* natural elements, and montane rivers in particular, empower the protagonist and narrator, Ernesto, to fight against the oppression of Indigenous groups. Rivers also become tools for self-identification, since—living between two worlds (the Indigenous and the Criollo) but not entirely belonging to either one—Ernesto opts for an Indigenous worldview, as opposed to the Criollo cosmovision of his father and mainstream Peruvian society. As will be seen in the next section with the presence of the Ranchería River in the novel *Palabrero*, in *Los ríos profundos*, rivers—and especially the majestic Apurímac (figure 4.1)—are also conceived as living beings whose dynamism (they seem to talk to Ernesto) contains agency that affects nearby humans (themselves part of the same natural network or web of life, rather than its masters).

In spite of not being of Indigenous descent, the semiautobiographical fourteen-year-old Ernesto rebels against the mistreatment suffered by Indigenous people at the hands of the landowners who exploit them. The bildungsroman delineates how, through introspection and with the help of his dreamlike communication with rivers and other natural elements, Ernesto learns to interpret the surrounding reality and to make ethical choices: eventually, unhappy with the outcomes of modernity and westernization in the Andean world, he chooses the side of subaltern Indigenous groups.

Through the re-creation of the Andean worldview, Arguedas then vindicates Quechua culture and its way of being in the world. This is particularly evident whenever the adolescent Ernesto leaves the boarding school to chat with Andean nature during his lonely walks. Unlike the Mestizo groups in the *chicherías* (chicha bars) of Huanupata or the Indigenous colonos in Patizamba, the natural world accepts Ernesto with open arms for who he is; in the protagonist's view, rivers and plants are the only ones who understand him, his only friends. This mentality responds to his early adoption of an Indigenous belief system that sees human beings as integrated with nature and with the universe as one. Toward the denouement of *Los ríos profundos*, after leaving Abancay, a concientious and idealistic Ernesto feels proud to have decided not to meet El Viejo (the representative of Criollo oppression against Indigenous people, but also of economic security for Ernesto), joining instead the Quechua people in their march, while he waits for his father to return.

The novel, as is well known, was Arguedas's way to protest the idealized and exoticized representation of Andean people in the Indigenista novels of his time, mostly written by Mestizos. As a man who grew up among Quechua people, he felt more informed and entitled to represent their world from within. With that goal in mind, the author Quechuanizes the Spanish language to better address the representation of Andean culture, all the while reaching a large readership. Both the author and his alterego, Ernesto, believe in the ethical and moral superiority of native people, and more important, in their capacity for self-emancipation. In fact,

FIGURE 4.1. Apurímac River in Peru. Photo courtesy Robert Bradley.

the novel has an implicit liberationist and revolutionary message that, to Arguedas's dismay, was ignored by most critics: even though toward novel's end, the
Indigenous colonos' uprising responds to their belief that a Catholic mass will help
the souls of their plague-ridden and recently deceased brethren reach salvation in
heaven, it is still an epic demonstration—along with the previous, failed rebellion

of the chicheras led by the indomitable and messianic leader Doña Felipa, who demanded the sharing of the salt with the native population—that they will not continue to remain submissive to the unjust status quo.

Bordering on what the West would perhaps perceive as a type of magical realism, the novel tries instead to depict Andean Indigenous culture and their struggle against epistemicide. In this sense, like neomaterialist theory, the novel proposes the existence of a harmonious continuity between humans, animals, rivers, and the natural world in general. Ernesto's communication with nature begins early in the first chapter, when his conversations with the ancient Inca stones that have survived in the buildings of the city of Cuzco. As happens when he is in the presence of Andean rivers, the protagonist senses the energy of these stones and feels that perhaps one day they too will start walking. It is apparent that for him these stones are not simply inorganic objects; rather, they are living, organic repositories of history willing to communicate their knowledge to humans. Moreover, those buildings, which blended Inca construction at the bottom and Spanish at the top, like the church bell built with Inca gold, embody the *mestizaje* that Arguedas saw as the essence of Peruvianness.

It is not surprising, then, that Arguedas chose to center the novel's titles on rivers, those deep rivers at the top of the Andean Mountains that preserve the deepest roots of native Peruvian culture. Tellingly, also, his first short-story collection was titled *Agua* (*Water*, 1935). Following native cosmogony, for Ernesto, the swelling of the Pachachaca, the Apurímac, and other montane rivers, breaching their banks, represent and announce the final awakening and liberation of Indigenous people. The agency of deep montane rivers purifies the protagonist's hurting soul, reminds us of the native roots of Peru, and announces, when flooding, the upcoming emancipation of the Quechua.

"IS IT POSSIBLE TO STEAL A RIVER?" WATER, EXTRACTIVISM, AND INDIGENOUS GENOCIDE

To continue with Latin American Indigenous worldviews and the agency of rivers, Philip Potdevin's novel *Palabrero*, published almost six decades after *Los ríos profundos*, opens with a question posed by the Wayuu Indigenous protagonist, Edelmiro Epiayú, to a Spanish lawyer who is taking care of him: "Is it possible to steal a river?"[9] The protagonist's denunciation of this theft and of the genocide of his own people appears early in the first paragraph of the novel: "Among the ways to exterminate a people, the most despicable is to drown it in its thirst. When the only source of water is removed, the connection with life, what can one do?"[10] As if foretelling the victorious denouement of the novel, Edelmiro evokes past collective struggles, literary and historical, against injustice, such as Antigone, Lope de Vega's *Fuenteovejuna*, and that of the Wayuu historical hero Juan Jacinto.

The Ranchería River, which in the novel is about to be displaced by a mining company, becomes synonymous with the physical survival of the Wayuu Indigenous community in the peninsula of La Guajira (also spelled La Wajira, as Potdevin does in the novel), in northeastern Colombia.[11] In hopes of improving his economic situation, Edelmiro had initially planned to work as a lawyer for a coal mining corporation that maintains a huge operation in the region. Yet not only is he rejected and discriminated against for being Wayuu, but he also realizes that coal mining is destroying his community.

The protagonist has a chance to observe the mine's huge *tajos* (gashes) that eviscerate Mother Earth in order to extract the coal, all of them with phosphorescent, green, contaminated water at the bottom. He then opts for fighting the mining corporation, instead of working for it, upon realizing the magnitude of the ecocide being committed, as well as the environmental racism his people are suffering: water contaminated with mercury and coal powder in the air are rendering both Indigenous children and adults sick with cancer, tuberculosis, conjunctivitis, diarrhea, vomit, and silicosis (which local doctors refuse to diagnose, as it would justify the protestors' demands). Eventually, Edelmiro sues the corporation, demands that they produce the permission for their coal extraction, and exposes the double standards of foreign shareholders, who seem satisfied with abiding by the Colombian environmental legislation, even though they are aware that it is much less demanding than the prevailing environmental legislations in their own countries. When the corporation attempts to buy the protagonist out by offering him a job as director of one of its foundations, he vehemently refuses it.

At one point, Edelmiro receives a message from the spirit of his deceased uncle and mentor, the *palabrero* (Wayuu type of lawyer or conflict mediator with great moral authority) Fulvio Epiayú, declaring him the chosen one who is to save the Ranchería River from being diverted, thus preventing its death as well as the decline of the Wayuu community. From that moment on, the protagonist sees himself and his community as the people in charge of defending Mother Nature from their "younger brothers," the "civilized," non-Indigenous people. Edelmiro and other Wayuu leaders see the diversion of the Ranchería River as a nail in the coffin of an Indigenous community already displaced and decimated by a gigantic coal mine run by a multinational corporation without scruples.

From the world of fiction, therefore, *Palabrero* exposes real-life, neoliberal appropriation and contamination of rivers and underground waters, as well as air pollution by coal dust and other pollutants, which is causing the deadly silicosis. In the novel, local miners, together with Wayuu children and elderly people, are dying in great numbers as a result of environmental racism. Initially, many Wayuu and their goats are being accidentally killed by a train with more than one hundred cars used for transporting thirty tons of coal a day. However, the protagonist later discovers that, for other desperate members of the Wayuu Indigenous community, including his beloved palabrero uncle Fulvio, this same train, a symbol of how modernity has failed the Wayuu, has simply become their chosen way to commit suicide: they can no longer cope with drought, disease, and dire poverty. Tellingly, Edelmiro sees this huge train as the shadow of the *Titanoboa cerrejonensis*, a fossil found in a coalmine in Cerrejón, La Guajira, in 2004, the largest fossil of a snake ever found in the world (thought to have been between forty-two and forty-nine feet long and weighed twenty-five hundred pounds).

The young Wayuu protagonist and hero of the novel is of uncertain origin, as he was found abandoned in a well when he was only a few hours old. This provides him with a sort of mythical and messianic aura, as if he were a new biblical Moses. That his adoptive mother immediately noticed his resemblance to the historical Wayuu hero Juan Jacinto, who led an uprising against the Spanish colonizers, hints at the baby's future heroism. Edelmiro narrates, in the first person, his long and arduous process of preparation, receiving first training as a lawyer at the Universidad de La Guajira in Riohacha (against his beloved uncle's wishes, who fears that he will lose his Wayuu identity) and then as an Indigenous palabrero. He basically becomes twice a lawyer with the goal of contesting political and judicial corruption in La Guajira, which has allowed the mining company to commit environmental crimes with impunity for almost three decades.

In this way, Potdevin places Indigenous, oral, ancestral knowledge on an equal footing with Western, written knowledge. Edelmiro's learning process includes, besides a university degree, access to ancestral Indigenous knowledge in Sierra Nevada de Santa Marta, at the very source of the Ranchería River. He is trained by the *mamu* (the Kankuamo version of a palabrero or shaman) Don Eleuterio, who reconnects the protagonist with Mother Earth (or the Universe). Don Eleuterio teaches him the native *Ley de Origen* (Law of Origin), which is described in the novel as wisdom emerging from Mother Earth and as the only law that the Wayuu and other Indigenous groups have, which is connected with the concept of seeking harmony with nature. Through meditation, Don Eleuterio helps him, over the course of more than half a year, to become one with the Ranchería River. Later, Don Eleuterio sends him to become familiar with the vallenato music of the city of Valledupar. Completing his professional and ethnic education is his sentimental education, thanks to his sexual experiences with both Indigenous and non-Indigenous or *alijuna* women. Eventually, this training turns into a sort of initiation, almost a rite of passage that opens his eyes to the immense challenge ahead of him: the "reborn" protagonist realizes that he will need all the collective help he can gather in order to face the powerful and heartless multinational coal mining corporation that is strangling his community and drying up its natural resources, including the most precious of all: water.

Potdevin, who received a law degree from the Universidad de San Buenaventura in 1984, takes advantage of his familiarity with the field to re-create the protagonist's legal challenge to a ruthless mining corporation that ignores international environmental regulations and manipulates legal processes through bribes. Guided by love for his community and culture, and surviving attempts at framing him for theft, mail threats, and even two assassination attempts (possibly carried out by landowners, smugglers, the owners of La Guajira, who have lucrative contracts with the corporation and fear that Edelmiro's protest may end up damaging their businesses), his resilience ultimately leads him to a victory that provides temporary relief for his people. With the help of like-minded, sympathetic people, he is eventually able to combat corruption and extortion.

In the happy denouement of the novel, Edelmiro is supported in a huge demonstration on World Water Day (March 22) not only by other Indigenous leaders and organizations but also by the entire country (including its opportunistic president, who seeks political capital with his participation) and even European activists and lawyers, the mining company is forced to close down its operations in La Guajira, and the Ranchería River is never displaced, thus assuring the Wayuu's cultural and physical survival. All things considered, the demonstration against the diversion of the Ranchería River ultimately becomes a national plebiscite in defense of Indigenous peoples.

Along the way, *Palabrero* delineates the intimate connection between Indigenous culture (not only of the Wayuu but also of other Indigenous groups in La Guajira, such as the Paraujanos and the Kusina) and the main source of water in the Guajira Peninsula: the Ranchería, its only river. Throughout the plot, we learn that water shortages and the displacement of Indigenous communities are accelerating the progressive loss of cultural identity, with young Wayuu moving to the cities or choosing to work for the main perpetrator of their community's demise: the mining company. To the protagonist's surprise, for instance, one day he realizes that his own long-time absent father has become one of the company's train conductors.

Although in the epilogue Potdevin acknowledges that all the characters except for the historical Wayuu leader Juan Jacinto are fictional, he is also careful to clarify that the coal mine does exist. In fact, the novel is modeled after the real fact that the mining company Cerrejón Limited (carbon extraction in the region began in 1985) indeed tried to divert the only river in La Guajira, the Ranchería, to mine the 550 million tons of coal under it. Potdevin, who is not of Wayuu ancestry, lived for almost five years in their land, La Guajira, where he was in both direct and indirect contact with the Wayuu and had a chance to familiarize himself with their customs, religious beliefs, and legends. In La Guajira, Potdevin witnessed firsthand their suffering and hopelessness, their struggle for survival, for recovering their pride in their heritage and ethnicity. The author also acknowledges reading sociological and anthropological studies by both Wayuu and non-Wayuu scholars Weildler Guerra, José Polo Acuña, and Michel Perrin, dealing with Wayuu history and culture, syncretic religious beliefs and mythology, and colonization by Europeans, all the way to today's oppression by non-Indigenous groups, Colombian authorities, and multinational corporations.

As the author clarified in a personal conversation, the novel was inspired by the horror of witnessing, while working in La Guajira, the collective self-extermination of an Indigenous community silenced by the media: the constant suicides by throwing themselves to the railways that have plagued the Wayuu community for the last three decades. He was also inspired by the 2012 real-life Wayuu opposition to the attempt by a mining company to divert twenty-five kilometers away the only river in La Guajira, the Ranchería—which the Wayuu consider sacred—in order to mine a huge amount of coal lying under it. As explained in the novel, also in real life the Ranchería has a natural system, an aquifer or sort of enormous sponge, to store water under it during the rainy season, thus preventing it to run dry out entirely during the dry seasons, from January through March and from June through August:

> The river is a living being, it wants to avoid overflowing, to damage the inhabitants of its banks, for that reason, it begins a loving act of giving, of receiving: it shares the water that it is carrying in excess at that moment with its deep bed, because it does not need it all to advance to his destination in Riohacha, where it will pour the sweet liquid onto the salty Caribbean. That way, therefore, it gives that excess water to the sponge, nature's gesture of generosity and love, of a harmonious pairing between river and sponge.[12]

The river's diversion, therefore, would have caused its disappearance for months at a time, since it could not resort to the ancient aquifer during dry seasons. In 2012, real-life local Indigenous communities demonstrated their awareness of the terrible consequences that such an ecocidal act would have had for the environment as well as for their people's physical and cultural survival. Still, the presence of the largest coal mine in Latin America and the tenth largest in the world continues to put at risk the survival of more than sixty thousand Indigenous people in the municipalities of Maicao, Barrancas, Hatonuevo, and Albania, among others, who continue to be displaced and made sick.

Palabrero includes additional sociopolitical criticism, all of it inspired by real-life facts. It denounces, for instance, the decades-long malnutrition and numerous deaths of Indigenous children due to respiratory infections, diarrhea, and diseases transmitted through food or water. It also bemoans the desecration of Indigenous family cemeteries and sacred lands (such as temples and "payment sites" for diseases, controlling death, etc.) by the mining corporation, as well as their frequent refusal to pay for a fair compensation. Thus, one of the Indigenous women living near the mine whom the protagonist meets describes the chilling scene of the destruction of one of the family cemeteries:

> When we thought they were finished, we heard a new haul of the machine. My mother cried out desperately when a new body came out of the earth: "It's my mom, baby, it's my mom!" And, indeed, it was the torn-apart body of my grandmother. I recognized it upon seeing the blanket with which we had buried her two or three

months earlier. We could not see this spectacle anymore and returned to the ranche-ria, our heart torn to pieces and rage in the blood.[13]

In addition, the novel celebrates the protagonist's victory against the opprobrious practice of changing the birth dates (they assign December 31 as the birth date of almost all the Wayuu) and Wayuu names to grotesque-sounding ones (such as Pescado [fish], Teléfono, Mariguano, or Chorizo) in their identity cards, and then taking all these identity cards away in order to manipulate political elections. Logi-cally, all these affronts were affecting the Wayuu's identitarian self-esteem, dignity, and ancestral ethnic pride.

Besides coal burning's irrefutable contribution to climate change, *Palabrero* criticizes yet another dark side of the relative economic wealth produced by mining for a few people in the region: social pathologies such as child prostitu-tion, theft, and drug trafficking. The novel also exposes generalized Colombian racism, when Edelmiro is discriminated against as a lawyer, despite his brilliance, only because of his native blood: "This nation is more racist than the United States or European neo-Nazis. It does not like indigenous people, it despises the Wayuu, the Uwa, the Aruhaco, the Kogi, the Nasa, the Wiwa, the Nukak-Makuk. They do not want to see their daughters married to someone who can generate a jump back for the race."[14] He goes as far as to compare it to a caste system. As the novel progresses, the reader finds out that this day-to-day racism is translated into a systemic environmental racism that is openly tolerated, if not encouraged, by the Colombian government and by foreign shareholders.

Potdevin has expressed in several interviews his interest in providing a voice for the subaltern Wayuu people and in improving their pride and self-esteem through this novel. Thus, in an interview with the Colombian journal *El Heraldo*, he states:

> The novel gives voice to Wayuu pride, which has long been affected by various social, political, cultural, and administrative factors. The Wayuu have a glorious past that has largely been forgotten and is unknown to new generations. *Palabrero*—which takes the title from the role played by the *palabreros*, bringing the word back and forth in order to reach an agreement between the parties—is a cry, a literary one of course, for the Wayuu to be heard, dignified, and given back their autonomy, values, and rights. All of this belongs to them, as they have lived in La Guajira Peninsula much longer that the alijuna.[15]

Thus, through his fiction, the author denounces the Colombian government's complicity with mining corporations and its indifference toward the long-lived Wayuu suffering in La Guajira. He also encourages the Wayuu community to fol-low the example of their historical hero Juan Jacinto, to wake up and use the peace-ful but powerful word, as the palabreros do, in their struggle for survival.

The implicit message of the novel is that the Wayuu should try to recover their greatness, strength, pride, and tradition of bravely facing adversity by imitating the pride of the indomitable Juan Jacinto. It even suggests how to do it when Edelmiro, emulating his uncle Fulvio, goes to the Sierra Nevada de Santa Marta to seek advice from a mamu from a different Indigenous group: "He [Edelmiro's uncle] did what many Wayuu palabreros have never done or if they do it, they do not admit it: leaving La Wajira and looking for affinities with our blood cousins, the inhabitants of the Sierra. The Wayuu will be strong if we join the communities of our brothers. The peninsula begins in the Sierra Nevada."[16] Therefore, learning from the heroes of the past and seeking unity and solidarity among all the Indigenous groups in La Guajira Peninsula are presented as the path to survival.

As a warning, the signs of impending sociocultural doom for Indigenous communities are apparent throughout the text. For instance, in the Sierra Edelmiro finds out that the last speaker of the Kankuamo language died three years earlier. Soon after, he learns that entire forests have been cut down to use the land for coca fields that will supply the narcos. He is also appalled to see that even the traditional palabreros of his region now wear fake Ray-ban sunglasses: "In short, its appearance was a sample of the overwhelming cultural syncretism, of how far Westernization destroys the original cultures of America."[17] In other words, traditional Indigenous culture is dying before Edelmiro's eyes.

The title of the novel, *Palabrero*, is taken from the profession of key members of Wayuu society: men who are in charge of solving interfamilial conflicts by going back and forth between the clans involved, taking "the word" in a sort of give and take, including offers for reparation. Known as *Putchipuu* in the Wayuu culture, this person, a sort of Indigenous lawyer and moral authority, mediates in conflicts using the word as his only tool for justice. In the novel, we learn about the mythical origin of the palabrero profession: the Utta bird was the first palabrero among the Wayuu people and the one who taught them how to live in harmony. As the novel explains, some palabreros consult with elderly Wayuu women about preestablished norms by asking how similar conflicts were resolved in the past.

Potdevin dignifies this ethnic community in *Palabrero* by bringing to life its glorious past and treating its culture with respect in the novel. Thus, the narration is intercalated—in a way reminiscent of Mario Vargas Llosa's *El hablador* (*The Storyteller*, 1987)—with short chapters or passages devoted to the Wayuu's origin myths, belief system (we learn, for instance, that the Wayuu die twice,[18] as announced in the epigraph), and the story of Juan Jacinto, the little-known, historical, Wayuu hero who fought the Spaniards during colonial times. The author found the story of this Wayuu hero in a historical study by José Trinidad Polo Acuña. According to Potdevin, most Wayuu and, by extension, most Colombians are unaware of the story of this real-life wajiro cacique, Juan Jacinto, who in 1769— ten years before José Antonio Galán's Insurrection of the Comuneros against the Spaniards—expelled the Capuchino orders who were forcing Indigenous women to live with them, taking Indigenous children to orphanages, and even abusing them. Juan Jacinto then led seven hundred warriors and twelve other caciques (many of them his relatives) to a seven-year uprising against the Spanish colonizers in the northern section of La Guajira Peninsula, north of Riohacha. The historical hero functions as an eighteenth-century döppleganger and role model for the protagonist in his struggle for the survival of the Wayuu and Mother Earth. Potdevin establishes, therefore, a parallel between the historical Wayuu leader and Edelmiro Epiayú, his fictional protagonist, underscoring along the way the centuries-old, continuous Indigenous struggle against the abuses committed by the alijuna or non-Indigenous people in this region. Likewise, the Wayuu in the novel learn not to make the same mistake that brought down their ancestors' rebellion: just as the Spaniards managed to divide and conquer by compensating one by one the Wayuu leaders who supported Juan Jacinto for their losses, the mining company is now attempting to buy out with gifts several Indigenous communities or *rancherías* one by one.

The committed nature of this novel is apparent once the fictional account is compared with the real-life events that inspired it. Thus, Alejandra Correa, in her 2017 article "Al pueblo wayúu se le agota el tiempo" (The Wayuu People Are Running out of Time), questions the pro-environment promises made by Lina Echeverri, vice president of communications and public affairs for the coal mining company Cerrejón Limited, upon the company's improvement in a "best practices" ranking:

> We have been in La Guajira for more than three decades under the premise of building transparent relationships with our stakeholders, being a good employer, complying with the law, and applying the highest social and environmental standards. Concomitantly with positioning ourselves as a leading producer and exporter of coal worldwide, we are focused on continuing to be a key ally for the progress and sustainable development of La Guajira, strengthening the capacities among its authorities and communities to lead the social transformations of its territory.[19]

As *Palabrero* makes clear, however, the mining company is not known for its efforts at sustainability. Along these lines, Correa points out that, even though the company continues to claim a commitment with the Wayuu and the local environment, arguing that they mostly use contaminated rainwater, in reality it is precisely because of the coal company that the contaminated rainwater falls all over La Guajira, adversely affecting the Wayuu's health and agriculture.

Similarly, *Palabrero* denounces another case of environmental injustice, the fact that while local Indigenous people have to withstand constant drought, it never affects the mining company: "Water is never scarce. They do not know what a drought is, a summer, a water cut, let alone an energy cut. The creeks Bruno, Palomino, the Ranchería River are its inexhaustible source. For them, there will always be water."[20] Indeed, more than anything else, the theft of the scarce water in the region and the displacement of the Ranchería River are the main complaints by the Wayuu in the novel.

Correa likewise questions the environmental plans announced by León Teicher Grauman, former president of Cerrejón Limited, once coal mining activities end:

> When the mining activity concludes in the future, we would like to see this laboratory of tropical dry forest developed by Cerrejón, turned into a fundamental element within the main ecological structure of La Guajira, which guarantees the provision of ecosystem services and sustains the prosperity of this noble department.[21]

In Correa's opinion, by the time mining activities cease in 2034, as planned, the land will be ecologically unrecoverable; and even if it were to be rehabilitated, it would still be privatized rather than returned to the Wayuu. Potdevin's novel also addresses these purported recovery plans and exposes the small, already rehabilitated areas as a mere publicity stunt:

> "It's for the photo, Édel, do you understand?"
>
> "Nope."
>
> "When visitors or journalists arrive, or the president or ministers, they take them to that area of rehabilitation, they show them the handful of hectares they have rehabilitated, they express their commitment; of course, these guys, if they are naive, are happy, proud to see a responsible handling of the environment."[22]

Correa also laments the fact that the official reports about mortality in La Guajira are not trustworthy. Also contesting official statistics, when in the novel Edelmiro visits two Indigenous communities near the giant coal mine, Provincial and Cerro Hatonuevo, the first scene he observes is one with three weeping Wayuu women in mourning clothes and carrying the dead bodies of their children. Likewise, although the Colombian state recognizes the existence of 270,413 Wayuu, in reality there are 800,000 Wayuu (Correa 2017). This fact is also twice reflected in *Palabrero*. First when a character named Lucho uncovers the ruse:

> But there have been, and continue to exist, more subtle ways to prevent prior consultation. One of them is to ignore the existence of indigenous communities in the territory. Remember I mentioned that the first step is to request a certification from the Ministry of the Interior; you will be surprised, but in many cases the Ministry itself has denied that existence, arguing that there are peasants or settlers but not indigenous people as such.[23]

And later, Edelmiro finds evidence of the way in which the corporation has ignored national and international environmental norms, as well as the laws for the protection of Indigenous communities: "We found that the Ministry of the Interior had certified, in an absurd way, that there were no Wayuu settlements on the banks of the Ranchería River, so that the entire consultation process could be omitted on paper."[24] Fiction is, therefore, not that far from the sad reality of La Guajira.

In a similar vein, the protagonist of *Palabrero* reminds us that the Colombian constitution of 1883 gave Indigenous communities the status of underage children, incapable of deciding for themselves. Only the new constitution of 1991 changed that status, giving them their constitutional rights. Don Eleuterio, the mamu from the Sierra Nevada, advises the protagonist to take advantage, as a lawyer, of these new protections for Indigenous peoples, even if they are being currently ignored. This is indeed the strength of the "word" that is recurrently evoked in the novel: Edelmiro must use both his Indigenous knowledge and his training as a lawyer to make sure that the written word in the new constitution is respected and applied against the existential threat that has been imposed on his people. In addition, Indigenous groups must convince the government that native law must prevail within the black line where they live.

Echoing the novel's denunciation, according to Correa, by 2016 the coal corporation had diverted more than seventeen bodies of water and had already begun studies to divert one of the most important creeks in northern La Guajira: the Bruno Creek. In the novel, by contrast, the diversion of the Bruno and Palomino Creeks has already taken place: "That the Ministry of the Environment had not established any type of study for the diversion of the Bruno and Palomino Creeks, that these had already been intervened and modified in their original channels and that none of them had prior consultation processes."[25] In real life, as a result of the plan to divert the Bruno Creek, local Indigenous communities (more than twelve thousand Indigenous people live in the area) collected signatures against the diversion, an action that received national and international attention. Yet Cerrejón continues planning new diversions of bodies of water to take advantage of the thirty-five million tons of coal sitting under the Bruno Creek. This is particularly problematic because, as Edelmiro points out in the novel, the rainy season, which used to be in April and September, has virtually disappeared as a result of climate change.

The connection between the water of the Ranchería River and Wayuu identity is a leitmotif throughout *Palabrero*. Thus, at one point Edelmiro travels to the river mouth of the Ranchería and suffers upon realizing how polluted his sacred river has become:

> The waters looked dirty, stagnant, infested with garbage on both banks; the stench that I felt breaking my bones as soon as I left the house came largely from this place. The river is wounded, I thought, as if its sinuous body was a single great purulent wound. It is the only one in the Wajira.[26]

The personification of the river (it is "wounded") in this passage once again points at the different ways in which Indigenous groups view the natural world.

In the final analysis, the official and unofficial uses of water as a mining resource or its consideration as an obstacle (i.e., when the corporation plans to change the course of rivers and other streams to mine the land under it) bring us to the concept of *biopolitics*, coined by Michel Foucault in *The History of Sexuality* and elaborated in another of his books, *The Birth of Biopolitics*. In this last book, Foucault, somewhat ambiguously, defines biopolitics as "the attempt, starting from the eighteenth century, to rationalize the problems posed to governmental practice by phenomena characteristic of a set of living beings forming a population: health, hygiene, birthrate, life expectancy, race . . ." (2008, 317). Later, Giorgio Agamben would define it as "the growing inclusion of man's natural life in the mechanisms and calculations of power" (1998, 119). Altogether, most understandings of the term address the phenomenon of seeing human life (population, administration of life) as a key element of political problems, as a tool for power. Likewise, another concept coined by Foucault in *The History of Sexuality*, that of biopower, addresses how nation-states resort to sociopolitical power to control human bodies and lives:

> During the classical period, there was a rapid development of various disciplines— universities, secondary schools, barracks, workshops; there was also the emergence, in the field of political practices and economic observation, of the problems of birth- rate, longevity, public health, housing, and migration. Hence there was an explosion of numerous and diverse techniques for achieving the subjugation of bodies and the control of populations, marking the beginning of an era of "biopower." (1978, 140)

These disciplinary institutions and mechanisms are, therefore, used to regulate large human groups or populations. In turn, Achille Mbembe (2016) defines bio- power as "that domain of life over which power has taken control" (12), which "appears to function through dividing people into those who must live and those who must die" (16).

It is this last meaning of the term that we find more useful for the analysis of *Palabrero*. With its tolerance—or even encouragement—toward the ecocide and genocide being committed in La Guajira by the mining corporation and its aco- lytes, the Colombian government actually decides who will live (alijunas or non- Indigenous people, including foreigners, with their incessant appetite for economic enrichment) and who will die (the Wayuu and other Indigenous groups in the region who can no longer withstand the drought and misery). *Palabrero*, therefore, delineates the connection between institutional and day-to-day racism and the environmental racism that is impoverishing the Wayuu and making them sick to the point where many have lost any desire to continue living. Legal and extralegal decisions about who owns the water and how it can be used are a form of biopower that ultimately deems Indigenous life unworthy or disposable. The Colombian state's management of health and food hygiene—or lack thereof—together with its permissiveness toward Indigenous deaths points precisely toward this material aspect of power that the novel exposes: society is not only socioculturally but also materially constructed and, as a result, water management is power over life.

In neomaterialist terms, the disrespect for the agency of water as a living being (as it is repeatedly defined by Indigenous characters throughout the novel) in La Guajira, and of the Ranchería River in particular, ends ups affecting human lives. The dynamic quality of the river, a sacred living being for the Wayuu, affects the human beings around it, who have to contend with its agency. In this sense, Jane

Bennett explains, "Thing-power is a force exercised by that which is not specifically human (or even organic) upon humans" (2004, 351). Bennett adds that thing-power materialism is a (necessarily speculative) onto-theory that presumes that matter has an inclination to make connections and form networks of relations with varying degrees of stability. Here, then, is an affinity between thing-power materialism and ecological thinking: both advocate the cultivation of an enhanced sense of the extent to which all things are spun together in a dense web, and both warn of the self-destructive character of human actions that are reckless with regard to the other nodes of the web (Bennett 2004, 354).

Like humans, therefore, nonhumans, including things, form networks by making connections with their surroundings. In the novel, the Ranchería River and its aquifer are the abused thing-power with the agentive force (it is not simply a cause) to affect the life or death of the Wayuu community. The river's survival equates the survival of the Indigenous community and empowers it to continue their struggle against neoliberal internal colonialism. The Ranchería ultimately helps the protagonist and his social group to engender subjectivity.

CONCLUSION

Both novels, Arguedas's *Los ríos profundos* and Potdevin's *Palabrero*, suggest that rivers and, by extension, the natural world are more than inorganic objects for their Indigenous or culturally Indigenous characters. Albeit written by non-Indigenous writers, both novels try to incorporate an Indigenous perspective of nature, which provides agency (or at least symbolic agency) to rivers as dynamic living beings. Rivers talk to the protagonists, calm them down, remind them of the deep native roots of national culture, or hold the key for the survival of Indigenous communities. Whether guided by cultural or existential survival, Indigenous characters also conceive of rivers in Peru and Colombia as a central part of their cultural and national identity. Their flooding may symbolize the future liberation of an oppressed people; their pollution or disappearance, the end of an Indigenous culture. A colonial bridge over it may symbolize the mestizaje of Spanish and Indigenous bloods. That is perhaps the essence of their sacred nature.

REFERENCES

Agamben, Giorgio. 1998. *Homo Sacer: Sovereign Power and Bare Life*. Translated by Daniel Heller-Roazen, edited by Werner Hamacher and David E. Wellbery. Stanford: Stanford University Press.

Arguedas, José María. 1935. *Agua*. Lima: Compañía de Impresiones y Publicidad.

———. (1958) 2004. *Los ríos profundos*. Edited by Ricardo González Vigil. Madrid: Cátedra.

Bennett, Jane. 2004. "The Force of Things: Steps toward an Ecology of Matter." *Political Theory* 32 (3): 347–72.

Coole, Diane. 2010. "The Inertia of Matter." In *New Materialisms: Ontology, Agency, and Politics*, edited by Diana Coole and Samantha Frost, 92–115. Durham, NC: Duke University Press.

Coole, Diana, and Samantha Frost, eds. *New Materialisms: Ontology, Agency, and Politics*. Durham, NC: Duke University Press.

Correa, Alejandra. 2017. "Al pueblo wayúu se le agota el tiempo." *Desde abajo*, November 25, 2017. www.desdeabajo.info/colombia/item/32972-al-pueblo-wayuu-se-le -agota-el-tiempo.html.

de la Cadena, Marisol. 2015. *Earth Beings: Ecologies of Practice across Andean Worlds*. Durham, NC: Duke University Press.

Foucault, Michel. 1978. *The History of Sexuality*. Translated by Robert Hurley. New York: Pantheon Books.

———. 2008. *The Birth of Biopolitics: Lectures at the Collège de France, 1978–1979*. Translated by Graham Burchell, edited by Michel Senellart. New York: Picador.

Frost, Samantha, and Diana Coole. 2010. "Introducing the New Materialisms." In *New Materialisms: Ontology, Agency, and Politics*, 1–43. Durham, NC: Duke University Press.

González Vigil, Ricardo. 2004. Introduction to *Los ríos profundos* by José María Arguedas, 9–108. Madrid: Cátedra.

Mbembe, Achille. 2003. "Necropolitics." *Public Culture* 15 (1): 11–40.

Potdevin, Philip. 2016. *Palabrero*. Bogotá: Intermedio.

Vargas Llosa, Mario. 1989. *The Storyteller*. Translated by Helen Lane. New York: Farrar Straus Giroux.

NOTES

1. In her book, de la Cadena explains that in Andean cosmovision "earth-beings"—or *tirakuna* in Quechua—are entities that blur the lines between human and nature (rivers, mountains, lagoons, etc.) as they have agency and the power to influence the course of everyday life decisions for both human and nonhuman entities.

2. In the novel, *las chicheras* (criollas) and *los colonos* (Indigenous people) are the two groups that represent the Andean Quechuan peoples. Two revolts occur in the novel. The first one takes place in chapter 7 and is led by Doña Felipa and other chicheras to avenge the abuses of the hacendados in the region and take back the salt that was stolen from the Quechua communities. Although this first revolt is unsuccessful, it triggers Ernesto into action and induces him to position himself on the side of the Indigenous cause. The second revolt, in chapter 10, is led by the colonos, who seek salvation through prayer and demand that Father Linares conduct mass after a disease has spread through Abancay and the nearby towns. This second revolt is successful and reinforces Ernesto's loyalties as he intercedes for the colonos and urges Father Linares to pray for them.

3. "Arguedas connota la profundidad—las sólidas, ancestrales, raíces, matrices de la identidad nacional del Perú—de la cultura andina, en contraposición al carácter sobreimpuesto—violencia de la dominación, actitud de dependencia de una metrópolis extranjera, desprecio y marginación de las raíces autóctonas—de una cultura occidental y cosmopolita a espaldas de legado histórico milenario del Perú" (González Vigil 2004, 75).

4. "Alude a que las construcciones sociales y culturales realmente sólidas y auténticas deben nutrirse (conforme hacen los ríos de la costa que descienden de los ríos serranos) de elementos andinos, indígenas, siendo ello un mestizaje fecundo y sintetizador del Perú" (González Vigil 2004, 76).

5. "Yo no sabía si amaba más al puente o al río. Pero ambos despejaban mi alma, la inundaban de fortaleza y de heroicos sueños. Se borraban de mi mente todas las imágenes plañideras, las dudas y los malos recuerdos" (*Los ríos profundos*, 232).

6. "Recibiría la corriente poderosa y triste que golpea a los niños, cuando deben enfrentarse solos a un mundo cargado de monstruos y de fuego y de grandes ríos que cantan con la música más hermosa al chocar contra las piedras y las islas" (*Los ríos profundos*, 196).

7. "¡No te olvides, mi pequeño, no te olvides!

Cerro blanco, hazlo volver;

Agua de la montaña, manantial de la pampa que nunca muera de sed.

Halcón, cárgalo en tus alas y hazlo volver.

Inmensa nieve, padre de la nieve, no lo hieras en el camino.

Mal viento, no lo toques.

Lluvia de tormenta, no lo alcances.

¡No, precipicio, atroz precipicio, no lo sorprendas!

¡Hijo mío, has de volver, has de volver!" (*Los ríos profundos*, 202–3).

In the novel, Arguedas provides both the Quechua and Spanish versions of the song next to each other, but writes the Quechua original version on the left in order to emphasize its true Quechua origins and the importance of language.

8. "Arguedas inserta elementos de la cultura andina (mentalidad mítico-mágica con sincretismo cristiano, quechuización del español, tradición oral ligada a la música y la danza) dentro de formas culturales occidentales (incluyendo entre ellas el propio género de la novela), en un grado de transculturación mayor que el logrado no sólo por los restantes autores vinculables a la corriente *indigenista* o *neo-indigenista*, sino por otros grandes cultores del *realismo maravilloso* hispanoamericano" (González Vigil 2004, 12).

9. "—¿Es posible robarse un río?" (González Vigil 2004, 13).

10. "—Entre las formas de exterminar un pueblo la más indigna es ahogarlo en su sed. Cuando se quita la única fuente de agua, la conexión con la vida, ¿qué se hace?" (González Vigil 2004, 13).

11. Philip Potdevin Segura, a novelist, essayist, and poet, was born in Cali (Valle del Cauca, Colombia) in 1958. Among many other literary awards, he won, in 1994, the Premio Nacional de Novela de Colcultura with the novel *Metatrón*. He has taught literary creation at the Universidad Central and also works as a literary coach of new writers. He has published the novels *Metatrón* (1995), *Mar de la tranquilidad* (1997), *La otomana* (2005), *En esta borrasca formidable* (2014), *Palabrero* (2016), *Y adentro, la caldera* (2018), and *La sembradora de cuerpos* (2019). He has also published the short-story collections *Magister Ludi y otros relatos* (1994), *Estragos de la lujuria (y sus remedios)* (1996, extended edition in 2011), *Solicitación en confesión* (2015), *Los juegos del retorno* (2017), and *Quinteto*. In addition, he has published the poetry collections *Cantos de Saxo* (1994), *Horologium* (1995), *Mesteres de Circe (Opus Magnum)* (1996), *25 Haikus* (1997), *Cánticos de éxtasis* (1997), and *Salto desde el acantilado* (2001). He has also published the literary translations *El ritmo de la vida y otros ensayos* by Alice Meynell (2001), *Voces Áureas*, by Pitágoras de Samos (2002), and *Oración por la dignidad humana* (2002).

12. "El río es un ser vivo, quiere evitar desbordarse, hacer daños a los habitantes de sus riberas, por ello, comienza un amorososo acto de dar, de recibir: el agua que lleva en ese momento en exceso la comparte con su lecho profundo, pues no la necesita toda para avanzar a su destino en Riohacha, donde verterá el dulce líquido en el salobre Caribe. Así, entonces, entrega a la esponja ese exceso de agua, en un gesto de generosidad y amor de la naturaleza, de armonioso maridaje entre río y esponja" (*Palabrero*, 96).

13. "Cuando pensábamos que habían terminado, un nuevo lance de la máquina, mi madre gritó desesperada, cuando salió un nuevo cuerpo de entre la tierra, '¡Es mi mamita, hijita, es mi mamita!'

y, en efecto, era el cuerpo despedazado de mi abuela, lo reconocí por la manta con a que la habíamos enterrado hacía dos o tres meses, no pudimos más presenciar ese espectáculo, nos devolvimos a la ranchería, el corazón vuelto pedazos y la rabia en la sangre" (*Palabrero*, 159).

14. "Esta nación es más racista que Estados Unidos o los neo-Nazis europeos. No gusta lo indígena, se desprecia al wayuu, al uwa, al aruhaco, al kogi, al nasa, al wiwa, al nukak-makuk. A sus hijas no las quieren ver casadas con alguien que pueda generar un salto atrás en la raza" (*Palabrero*, 29).

15. "La novela da voz al orgullo wayuu, tan afectado desde hace mucho tiempo por varios factores sociales, políticos, culturales y administrativos. El wayuu tiene un pasado glorioso que en gran parte se ha olvidado y es desconocido para las nuevas generaciones. Palabrero—que toma por título el rol principal que ejercen los palabreros para llevar y traer la palabra hasta lograr un acuerdo entre las partes—es un grito, literario por supuesto, para que el wayuu sea escuchado, dignificado y restablecido en su autonomía, en sus valores, en su derecho; por todo lo que le corresponde, por haber estado en la península de La Guajira mucho antes de los alijuna."

16. "Hizo lo que muchos palabreros wayuu nunca han hecho o si lo hacen no lo confiesan: salir de la Wajira y buscar sus nexos con nuestros primos de sangre, los habitantes de la Sierra. Los Wayuu seremos fuertes si nos unimos a los pueblos hermanos. La península comienza en la Sierrra Nevada" (*Palabrero*, 59).

17. "En fin, su apariencia era una muestra del avasallador sincretismo cultural, de hasta qué punto la occidentalización destruye las culturas originarias de América" (*Palabrero*, 135).

18. The Wayuu believe that they die twice. When a Wayuu dies on Earth, first he or she becomes a *yoluja*, a spirit of the dead that can communicate with the living, often through dreams. At one point, they leave for Jepira through the path of the dead Indians, the Land of the Dead, in Cape Vela, by the Caribbean Sea, where they become *juya* or rain, dying again and being buried for a second time.

19. "Hemos estado por más de tres décadas en La Guajira bajo la premisa de construir relaciones transparentes con nuestros grupos de interés, ser un buen empleador, cumplir con la ley y aplicar los más altos estándares sociales y ambientales. Al tiempo de posicionarnos como un productor y exportador de carbón líder a nivel mundial, estamos enfocados en continuar siendo un aliado clave para el progreso y desarrollo sostenible de La Guajira, fortaleciendo las capacidades entre sus autoridades y comunidades para que lideren las transformaciones sociales de su territorio."

20. "—Nunca escasea el agua. No saben lo que es una sequía, un verano, un corte de agua, ni mucho menos de energía. Los arroyos Bruno, Palomino, el río Ranchería son su fuente inagotable. Para ellos siempre habrá agua" (*Palabrero*, 150).

21. "Cuando la actividad minera concluya en el futuro, quisiéramos ver este laboratorio de bosque seco tropical desarrollado por Cerrejón, convertido en elemento fundamental dentro de la estructura ecológica principal de La Guajira, que garantice la provisión de servicios ecosistémicos y sustente la prosperidad de este noble departamento."

22. "—Es para la foto, Édel, ¿entiendes?

—No.

—Cuando llegan visitantes o periodistas, o el presidente o ministros, los llevan a esa zona de rehabilitación, les muestran el puñado de hectáreas que han rehabilitado, manifiestan así su compromiso; por supuesto, los manes, cuando son ingenuos, quedan felices, orgullosos de ver un manejo responsable del medio ambiente" (*Palabrero*, 155).

23. "—Pero ha habido, y sigue habiendo, formas más sutiles de impedir la consulta previa. Una de ellas es desconocer la existencia de comunidades indígenas en el territorio. Recuerden que mencioné que el primer paso es solicitar al Ministerio del Interior una certificación; ustedes se sorprenderán, pero en muchos casos el mismo Ministerio ha negado dicha existencia aduciendo que allí hay campesinos o colonos y no indígenas como tal" (*Palabrero*, 259).

24. "Encontramos que el Ministerio del Interior certificó, de manera absurda, que en las riberas del río Ranchería no había asentamientos wayuu, con lo cual se podía, en el papel, omitir todo el proceso de consulta" (*Palabrero*, 276).

25. "Que en el Ministerio de Medio Ambiente no se había radicado ningún tipo de estudio para la desviación de los arroyos Bruno y Palomino, que estos ya habían sido intervenidos y modificados en sus cauces originales y que en ninguno se realizaron procesos de consulta previa" (*Palabrero*, 276).

26. "Las aguas se veían sucias, estancadas, infestadas de basuras en ambas orillas; la fetidez que sentí penetrarme los huesos tan pronto salí de la casa provenía en gran parte de este sitio. El río está herido, pensé, como si su sinuoso cuerpo fuera una sola gran llaga purulenta. Es el único de la Wajira" (*Palabrero*, 57).

Water as the Medium of Measurement

Mapping Global Oceans in the Nineteenth and Twentieth Centuries

Penelope K. Hardy

While humans have always interacted with the sea, the ways in which they have imagined and thus defined it have changed significantly over the course of history. Beginning in the nineteenth century and continuing into the twentieth, scientific investigations of the ocean increasingly attempted to understand and map the ocean in three dimensions. While in the West this effort started earlier, with late-eighteenth- and early-nineteenth-century British investigation of global tidal patterns, or even with seventeenth-century efforts to map undercurrents in the Mediterranean, it reached its full flower by the mid-nineteenth century, as scientific investigators from Europe and the United States began to explore the ocean environment from aboard ships.[1] These in situ studies accessed the otherwise inaccessible ocean bottom using an expanding series of sounding technologies, and they resulted in a series of new attempts to represent the ocean in its many physical and biological dimensions in various maps, charts, and diagrams. The ocean bottom thus became a necessarily technologically mediated space that scientists and their partners defined as available for enlistment towards their commercial, disciplinary, and nationalistic goals. To accomplish this, the water that filled the ocean's basins became a tool for knowing and interpreting these invisible contours.

Reimagining the ocean as not just a highway or fishing ground but as a territory with physical features and knowable topography required invention and repurposing of technology, but it also meant a shift in understanding of both the ocean space and the water itself.[2] The ocean had certainly always had a third dimension in the human imagination, as people understood the risk of ships grounding on hidden shoals or foundering in open water, but this third dimension was unknown, and to a large degree considered unknowable. Early modern thinkers occasionally considered the ocean bottom, as Athanasius Kircher did in his 1664

Mundus Subterraneus, where he imagined the various seas linked through some network of subterranean passages through which water might move, thus explaining the tides. But measuring the ocean in order to establish its bounds in the wake of the Western scientific revolution, especially in its third dimension, involved the use of new tools, including ships that could manage the distances and depths involved, implements that could extend the human grasp miles beyond the sight of their wielders, engines that could handle the immense lengths of line necessary to reach the bottom, and eventually new technologies such as echo-sounding, which repurposed the water itself into a medium of measurement rather than just the substance filling the space to be measured.

As the ocean's bounds became delimited by numbers and details, as its topography became more clearly settled, its investigators imagined it as more and more land-like, and they began both to portray it in land-like ways and to enlist it for the uses they imagined for land. As the ocean gained mountains and plateaus, those plateaus could be imagined holding telegraph cables. As the ocean gained geological detail, those details could be enlisted for scientific arguments about terrestrial geology. As the ocean bottom was redefined as land, that land could be imagined as territory—useful, knowable, and even possessable. At the beginning of Western cartographic history, as I discuss below, understanding of the ocean shifted slowly from discrete bodies of water that divided the land to a global ocean which was navigable and thus connected the globe. The new, scientific reimagining of the nineteenth and twentieth centuries thus echoed the earlier shift; as the ocean bottom became known and land-like, it could connect continents telegraphically, be enlisted in global geological arguments, and extend terrestrial territorial claims. At the same time, however, the ocean bottom remained—and remains—unknown in significant ways. This fluid identity did not inhibit its enlistment for human purposes; indeed, the degree to which the human image of the reimagined ocean bottom retained fuzzy boundaries enhanced rather than degraded its usefulness, its suitability for repurposing. A little knowledge allowed investigators to make claims that appeared scientific, while retaining enough unknowns to engage the imagination.[3]

EMPTYING THE 2D SEA AS A SPACE FOR SCIENCE

As geographer Martin W. Lewis has argued, "The current taken-for-granted system of maritime spatial classification did not, in fact, emerge in broad outlines until the 1800s and did not assume its full-blown form until the twentieth century" (1999, 189). Earlier traditions imagined the ocean first as a world-encircling river, and later as a series of seas, usually understood and named by location and adjacency to terrestrial territories. Maps changed to reflect this shifting conception, though they often reflected the coastal nature of travel, even to distant ports, and maps intended for navigation highlighted the coastal landmarks encountered en route and reflected the commercial nature of these ventures, whether those involved exploitation of ocean resources such as fish or simply travel to distant markets where spices or other commodities could be purchased. Neither these nor the terrestrial world maps created by cartographers to depict the expanding known world portrayed the ocean's depths. Cartographers populated the vast, empty ocean basins with symbols of both the danger and mystery of the unknown: sea serpents and depictions of other marine beasts, half-described or imagined.

From the late fifteenth century, though, Europeans began to venture beyond their coast-hugging routes and braved the open ocean. To account for this new mode of travel, navigational charts needed to extend beyond coastal features and display the oceans as basins that connected destination ports. The lines between them and their cartographic cousins thus blurred as more of the world became navigable. By the time Magellan's ships circumnavigated the world in the second decade of the sixteenth century, routes between all known places could be charted (Rozwadowski 2018, 77). The resulting map, however, showed the ocean in two-dimensional outline, like a cut paper silhouette, perhaps recognizable by its distinctive profile, yet lacking in detail and emptied of the earlier imagined creatures.

As historian Helen Rozwadowski has argued, this "emptying" of the ocean left it open to scientific investigation of various kinds (2018, 85–86). Instruments to measure the elevation of the sun and stars allowed the calculation of latitude by the early sixteenth century, but the accurate measure of longitude remained elusive and the source of much scientific (and governmental) interest until the development of accurate and robust chronometers towards the end of the eighteenth century (Andrewes 1996). Aside from accounting for the ship's position, extended voyages meant a growing incidence of scurvy, and while the curative effects of citrus, for instance, were already known, this was little help on an extended voyage and still did not explain the etiology of the disease, so captains and their sponsoring governments continued to experiment through the eighteenth century (Rozwadowski 2018, 97). Finally, the ocean's effects reached even into major ports like London in the form of tides, and a growing global system of transportation and communication allowed an integrated effort to understand, chart, and predict them by the early nineteenth century (see Reidy 2008).

The involvement of governments in these efforts was neither incidental nor inconsequential. The opening of extensive global ocean travel both enabled and enacted the beginning of what is now recognized as global imperialism. Whether for commerce, for domination, or—usually—some combination thereof, government-sponsored voyages meant government interest in the safety of ships and their crews, thus the need to solve the problems of scurvy, longitude, and tides. The interest of these same governments in the natural history of their destinations is unsurprising; in addition to the safety and security of their new possessions or trade partners, this knowledge let them know what was exploitable and how (Portuondo 2009). By the nineteenth century, government patronage of commercial sea interests was well established in the West, even as sea voyaging itself, and the purposes for it, were changing as the Industrial Revolution transformed transportation and communications. These phenomena led to a popular "discovery" of the ocean in the nineteenth century, as a larger proportion of the population not only voyaged by sea, but had greater access to the seaside as terrestrial destination and site for recreation as well as social and intellectual pursuit (Rozwadowski 2005). The same combination of phenomena drew natural historians to the shore as well. Eventually, they went beyond it, using ever-larger vessels to investigate—and map—the conditions and inhabitants of the sea bottom further and further from shore.

MAPPING THE DYNAMIC OCEAN

One of the first efforts to collect such information synoptically came—appropriately enough—from a naval officer motivated by a nationalistic desire to see his country compete in the realms of both science and global commerce. Commerce had long proven a primary motivator for science, and in particular ocean science, as demonstrated by projects ranging from early modern developments in astronomy to the British effort to understand magnetism and the tides. As superintendent of the U.S. Navy's Depot of Charts and Instruments beginning in 1842, Matthew Fontaine Maury was responsible for the storage, calibration, and maintenance of the navy's navigational instruments and chronometers, as well as procurement of its charts.[4] His office was also responsible for the storage of ships' deck logs, in which the captain and officers recorded incidents of the ship's voyage, but also observations of the winds, weather, and sea conditions, which were tied to the specific latitudes and longitudes at which they were encountered. These were legal records, and so had to be retained, but in general the navy had no further use for them, so they did little but molder and take up space. Maury, however, quickly realized that the data contained within them collectively described climatic conditions across broad swaths of the planet's surface and through each season of the year (for more detail, see Hardy 2016).

Maury's staff harvested this data, which he soon supplemented by soliciting the assistance of deploying naval captains and later merchant mariners, giving them standardized forms on which to record specific kinds of data while underway (Burnett 2009, 194). Maury sold his project to the navy hierarchy as a domestication of chart-making—after all, was American reliance for charts upon Britain, with whom the country had been at war a mere generation earlier, not a national security issue?[5] Maury imagined representing the accumulated information graphically, envisioning a chart on which "the experiences of a thousand navigators" would guide the neophyte "as though he himself had already been that way a thousand times before" (Maury 1856, vii). With the wealth of data his agents were now gathering, Maury did create new charts, beginning with the Atlantic Ocean, but beyond the traditional navigational chart he envisioned a much broader series of representations of what he recognized as a global ocean-atmosphere system.

Collecting the information he wished to portray was not as easy as he had hoped, however. The old logs were soon mined out, and his fellow naval officers proved less than universally enthusiastic and conscientious about collecting new data, no doubt something they saw as simply an additional administrative task. Maury began communicating with captains directly, begging them to collect as much data as they could at every point on their routes, including by sounding in deep water.[6]

Sounding by line was an ancient technique for measuring depth whose technologies had changed little over the years. The basic premise involved a weight tied to a line and then thrown overboard. When the weight hit bottom, which an experienced navigator could feel through the line, the amount of line payed out revealed the water's depth. Traditionally, however, sailors had only bothered to do these measurements near the coast, or in other situations in which potential shallow water meant a threat of running aground. While a few deep-water measurements had been attempted—and recently even managed—they were not routine; the procedure required stopping a ship mid-ocean and holding it as steady as possible for the length of the measurement, an effort in which the average sailor saw little utility in a location where the bottom posed no risk to the ship.[7] Maury, however, wanted these measurements not for safety, but to further his understanding of the ocean in all its dimensions.

He also asked captains to report any errors found on published charts, especially vigias—shoals charted on slender evidence—many of which he believed did not actually exist. While initially he asked his fellow officers to gather data as "a great favor," that approach ensured neither the quantity nor the consistent quality of data he wanted.[8] To standardize the data, Maury devised an "abstract log," a blank form that would prompt the observer to record specific kinds of data at specific times. This log, which evolved into a set of forms in slightly different formats for different users, was itself a technology of data collection and standardization. Logs had long been kept as official records of a ship's voyage; Maury's efforts at standardization co-opted that tradition to record and gather specific kinds of data on a scale impossible for him to otherwise access.

By October 1843, he had convinced his immediate superior of his project's value, and he wrote up a few pages of "Suggestions for the Attention for the Home Squadron." These instructions detailed for ships' officers not only what data he wanted, but also which instruments already on board their vessels could be used to gather it, where and how best to set them up, and how often to take observations. Maury's allies in the scientific community petitioned the secretary of the navy to instruct U.S. naval vessels operating worldwide to participate.[9] His lieutenants got to work compiling a chart of the Atlantic from the materials on hand which Maury hoped to publish by the end of 1845, but compliance among naval captains was spotty at best and covered only those routes frequented by naval vessels.[10] While the data thus provided some new insights, they were not as extensive as Maury had hoped, they provided little ability to check information between ships, and they left large swaths of the oceans unmeasured.

Following up on earlier suggestions by Representative Stephen Mallory, chair of the House Naval Affairs Committee, that American merchants and whale-men might constitute a storehouse of information about the ocean, Maury began recruiting among these captains, and soon asked the hydrographic bureau chief to make an official request. In return, Maury imagined the better understanding of the winds and currents thus attained could allow captains to choose more effi-cient routes for their travels, saving them time and money while decreasing risk. His plan then was to furnish merchants with his blank forms; those who returned them filled with data at the end of their voyage would be furnished with a copy of his wind and current chart.[11] In July 1847 he was distributing copies of the first sheet of his wind and current chart. By November the other seven sheets of the Atlantic were in press.[12]

Maury's charts differed from previous ones because instead of simply displaying the ocean's surface and the land—the backdrop of a voyage—they dis-played as well the voyages of each ship that had contributed to the project, color coded by month and season of travel. Maury was attempting to map not just the ocean's contours, but its dynamic physicality as experienced at sea. In fact, Maury argued explicitly that this amounted to charting the accumulated experience of those who had traveled that path before, as he had previously hoped to do, thus making it accessible to both neophyte and veteran navigators.[13] Small, comet-shaped graphics indicated the direction and force of winds encountered, as well as their consistency. Currents were marked with a number indicating velocity in knots alongside an arrow indicating direction. With the charts, Maury published a volume of explanations and sailing directions—initially a slim pamphlet, but of greater heft with each new edition over the following years. These both explained how to read the charts and, later, analyzed specific example tracks sent in by Mau-ry's corresponding observers.

When the *W.H.D.C. Wright*, a bark out of Baltimore, followed Maury's direc-tions and reached Rio de Janeiro from the Virginia Capes in just thirty-eight days, then returned in thirty-seven, a seventeen-day improvement over the previous average in each direction, other captains took notice. Many soon proved willing to participate in Maury's data-gathering scheme, receiving blank logs before their voyages, filling them out while underway, and receiving copies of the new charts and sailing directions upon their return, allowing Maury to begin work on similar charts of the rest of the Pacific and Indian Oceans as well.[14]

In addition to this greater geographic coverage, Maury began to plan charts dis-playing more kinds of information. The original set of charts with their collected tracks—the Track Charts—would be known as Series A. They were followed by Trade Wind Charts (Series B); Pilot Charts, which showed prevailing winds in various seasons (Series C); Thermal Charts (Series D); Storm and Rain Charts (Series E); and Whale Charts (Series F), which displayed seasonal sightings by species, giving whalers what amounted to a graphical database from which to make hunting decisions and biologists a census of sightings and behavior.[15] These later series were even more innovative in their depiction of Maury's imagined ocean, and further from the traditional naval chart. The storm and rain charts, for instance, divided the ocean into a grid, within each square of which a smaller table of numbers explained average seasonal conditions as experienced by Maury's sources. The chart thus represented the experience of ocean travel rather than the geographical boundaries of a voyage, while the whale charts refilled the void that had been emptied of its fantastical inhabitants with a more biological—and commercial—understanding of the ocean's occupants.

CONSTRUCTING THE OCEAN'S THIRD DIMENSION

One of Maury's most novel representations of the ocean, though it would perhaps not be as immediately useful as he hoped the rest of his charts to be, was a chart of the North Atlantic Ocean basin published in 1853 (figure 5.1). While Maury would not yet use the word *bathymetric* to describe measurements of the ocean depths and the charts on which they were depicted, his 1853 image attempted to chart the depths not just near the shores, as would be necessary for safe navigation through shoaling water, but across the entire extent of the Atlantic. Maury recorded the depths of the minimal number of deep-sea soundings available, a number probably in the dozens at that point for the entire Atlantic. He displayed appropriate skepticism at the deepest of them; the contours they implied seemed excessively steep, and he recognized that as soundings got deeper sailors could no longer feel the moment the lead hit bottom through the line, and the line involved was now heavy enough to continue paying out, especially if caught in a current. Over the course of his career, Maury and his lieutenants would work out new means of determining the bottom of a sounding that relied upon timing the payout of the line and noting a slowdown when the lead was no longer falling through the water.

Once Maury charted these scattered soundings, he then filled in the contours of the ocean basin in soft curves that still imagined the ocean as a largely empty place. A shallower swath stretches down the middle of the basin, south and slightly west of the Azores, a hint of the Mid-Atlantic Ridge that would be revealed a century later, but Maury's imagined ocean bottom was largely flat, empty, and still. He quickly revised this as additional soundings became available, and he was able to attribute the inaccuracies of the deepest early soundings to strong bottom currents. Indeed, he increasingly imagined the bottom as mountainous and rugged terrain, an undiscovered country whose exploration would elucidate the physical phenomena—the waves, currents, and tides—of the ocean it held.

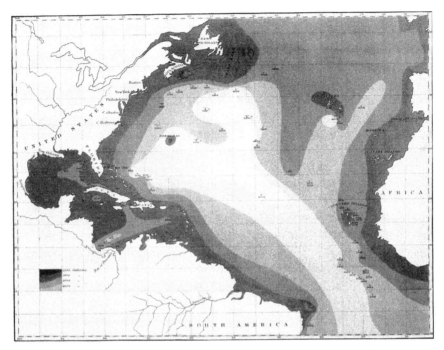

FIGURE 5.1. "Basin of the North Atlantic Ocean." Matthew Fontaine Maury, *Explanations and Sailing Directions to Accompany the Wind and Current Charts*, 5th ed. (1853). National Oceanic and Atmospheric Administration (NOAA) Photo Library.

Still, even when imagined as empty, these spaces held value—in some cases literal, commercial value—as Maury's support for the first efforts to lay a transatlantic submarine telegraph cable show. Soundings conducted by one of Maury's lieutenants across the Atlantic could be interpreted as a plateau of moderate depth, conveniently located along the great circle route between Newfoundland and Ireland. At the same time, one of his lieutenants had designed a new sounding tool which allowed the capture and return to the surface of small samples of bottom sediment.[16] These showed a landscape littered with the fragile shells of microscopic sea creatures, suggesting currents at the bottom were gentler than he had thought. Maury assured naval superiors and cable entrepreneurs alike that the plateau was as well suited for a cable as if it had been placed there for that purpose.[17]

DREDGING THE GLOBAL OCEAN

This new attention to the ocean's third dimension, and the use of biological data to elucidate its features, presaged the expansion of scientific investigation and accompanying redefinition of the global ocean in the second half of the century. The first and widest-ranging scientific expedition to study the global ocean in all its dimensions was the Challenger Expedition, a joint project of the British Royal

Navy and the Royal Society of the London.[18] Over the two-and-a-half-year period between December 1872 and May 1875, HMS *Challenger* circumnavigated the globe with a naval crew and a team of embarked civilian naturalists. While this was not the first effort to sound or even collect samples and specimens from the ocean bottom, the Challenger Expedition was the first whose primary purpose was the scientific study of the ocean itself.[19] The expedition's naturalists made significant zoological efforts; more than forty of the fifty official results volumes published in the twenty years after the expedition's return were on zoological topics. They also contributed to understanding ocean currents and firmly established the tradition of shipboard laboratory work. At the same time, the expedition significantly expanded the bathymetric understanding of the ocean that Maury's charts had begun.[20]

Biological questions served as the primary motive for the voyage, as British naturalists were inspired by and hoped to compete with recent Swedish work that had retrieved unusual biological specimens from 300 fathoms—about 550 meters—a depth then considered extreme.[21] Historian Rodolfo John Alaniz has argued that the publication of Charles Darwin's groundbreaking *Origin of Species* in 1859 provided further inspiration. Finding, or failing to find, the intermediate species whose existence Darwin had predicted in the seas provided an opportunity to test his assertions about the mechanisms of natural selection (Alaniz 2014, 228–29). However, the Challenger naturalists were also quite interested in geology. They and the ship's crew used sounding machines similar to those of Maury and his lieutenants but also deployed other tools adapted from fishers and oystermen, such as the trawl and the dredge. These were variations on a similar idea: a net held open by an iron bar or bars is dragged along the bottom at the end of a line by the movement of the ship above, gathering fauna, flora, and detritus in its path. Naturalists had begun to adapt the dredge by the mid-eighteenth century, but in the nineteenth its use from rowboats and other small craft to retrieve specimens of marine sea life was popularized by British naturalist Edward Forbes. Forbes understood this zoological work to have geological implications, noticing the similarity between the benthic fauna he retrieved and studied to similar fossil finds in now-terrestrial strata ashore. He predicted that a thorough understanding of the conditions under which these creatures lived—including pressure, temperature, and darkness—was key to understanding their natural history, from which naturalists could in turn extrapolate to understand the conditions under which the ancient terrestrial fossil beds had formed. Though Forbes died in 1854, others continued this work, including Scottish marine zoologist C. Wyville Thomson, who originally proposed the partnership that would lead to the Challenger trip.

During short, preliminary cruises of a few months each on naval survey ships, Thomson and his colleagues worked out which tools and implements were best suited to address their questions, the best practices for how to deploy them, and how a relationship between scientists and naval personnel would function aboard ship. With this background, the expedition proper could begin aboard *Challenger*. From 1872 to 1875, seven "scientifics," including Thomson, joined the ship's officers and crew for a scientific circumnavigation studying every sea except the Arctic.[22] A committee established by the Royal Society laid out the voyage's potential goals, including the investigation of the physical conditions of the deep sea, the chemical composition of sea water, the characteristics of bottom sediment, and the distribution of life.[23] In all, *Challenger* traveled 68,890 nautical miles, sampling the sea bottom at 362 points along the way, from the littoral to the deepest spot in the ocean (a location near the Mariana Trench now called Challenger Deep for just this reason).

The dredge was the key tool for sampling the bottom, and was soon modified to accommodate even smooth and rocky bottoms, from which it had originally returned empty. Dredging off the coast of Scotland during the preliminary cruises, the embarked naturalists brought up "a bluish-white tenacious mud" mixed with the microscopic shells of globigerina, a genus of planktonic marine foraminifera. The remains of these tiny animals had been found in abundance even by rudimentary earlier soundings of the North Atlantic bottom (Carpenter 1868). They had first been reported by Jacob Whitman Bailey, an American microscopist and professor at West Point, who examined samples obtained as part of Maury's program in the 1850s. Bailey reported that with the exception of only one sounding, "the bottom of the North Atlantic Ocean . . . from the depth of about 60 fathoms, to that of more than two miles (2000 fathoms), is literally nothing but a mass of microscopic shells."[24]

Now possessed of their own samples, however, Thomson and his colleagues could reimagine the ocean's place in geological history just as Forbes had suggested. In 1836, German microscopist Christian Ehrenberg had established through microscopic examination that the deposits of porous white limestone in Cretaceous strata in England and beyond—known as "the Chalk"—contained the remains of marine fossils, evidence that these layers had been laid down when the area was at the bottom of an ancient sea.[25] Now these sedimentary soundings from the Atlantic and beyond found these same creatures in the process of living, dying, and drifting into an accumulating layer at the bottom of the modern sea. This offered powerful support for uniformitarianism—the idea posited in the early 1830s by geologist Charles Lyell's (1831–33) influential volumes *Principles of Geology*, which argued that the earth had been shaped over an extremely long period of time via slow-moving processes, processes that were still acting at that time—but Thomson took it further.[26] This was not simply evidence of the same mechanism in operation today that had formed the Cretaceous Chalk in the past, he argued. Indeed, he claimed, "it is not only chalk which is being formed in the Atlantic, 'but *the* chalk, the chalk of the cretaceous period'" (Thomson, Carpenter, and Jeffreys 1873, 472, emphasis in the original). Other naturalists pushed back on the claim "that we might be regarded in a certain sense as still living in the cretaceous period." Thomson eventually changed his wording (after all, geological periodization is "thoroughly indefinite") but maintained, with Carpenter and Jeffreys, "that the balance of probability is greatly in favour of the chalk having been uninterruptedly forming over some parts of the area" that is now the Atlantic Ocean (Thomson, Carpenter, and Jeffreys 1873, 471–72). He thus used these sediments to reimagine not just the geographical space of Europe and the Atlantic, but their geological time, as well.

Based on the earlier American observations and their own on the preliminary cruises, the *Challenger* naturalists had set out expecting to find "a more or less universal chalk formation at the bottom of the ocean" (Buchanan 1919, 35). Yet as *Challenger* continued into deeper water, they noticed a shift from the "globigerina ooze" they had been retrieving to a red clay with no signs of foraminifera. The transition from organic to inorganic sediment was not abrupt but first passed through a transitional region of "gray ooze" at depths of between about twenty-two hundred and twenty-six hundred fathoms, where "the shells gradually lose their sharpness of outline, assume a kind of 'rotten' look and a brownish color, and become more and more mixed with a fine amorphous red-brown powder, which increases steadily in proportion until the lime has almost entirely disappeared" (Thomson 1878, 212–13). The naturalists thought it unlikely that the foraminifera that had thus far proved ubiquitous did not live in these areas, and the extremely fine texture of the remaining sediment made it unlikely that some sort of current swept them away. The gradual transition—with foraminifera shells in the intermediate area showing increasing degradation—instead suggested a chemical reaction.

Unlike earlier expeditions, whose specimens had frequently had to wait for analysis ashore, *Challenger's* onboard laboratory included chemical apparatus and microscopes. When expedition chemist John Buchanan subjected a sample of globigerina ooze to weak acid in the onboard chemical laboratory, the resulting product was a reddish mud (Thomson 1878, 215–17). This new theory, though, did not stand further testing; chemical reaction explained the disappearance of the globigerina, but when these creatures were captured in surface nets and subjected to the same chemical process, red clay did not result (Buchanan 1919, 34). Murray then suggested that the clay was in fact the result of decomposing pumice, expelled by volcanoes and left to drift globally until it finally sank and slowly decomposed. Under the microscope, the red clay proved to contain glassy feldspar and to lack quartz, evidence to support its volcanic origin (Buchanan 1919, 34).

As the ship proceeded into deeper water, the sediment samples showed another shift. Below three thousand fathoms' depth, the red clay began to accumulate the shells of radiolaria, a planktonic protozoa whose shells consist of silica instead of the calcium carbonate of the globigerina. Again, the transition was gradual, until eventually the red clay gave way to a siliceous "radiolarian ooze." This, too, reflected ancient geological conditions Ehrenberg had observed in Europe (Murray and

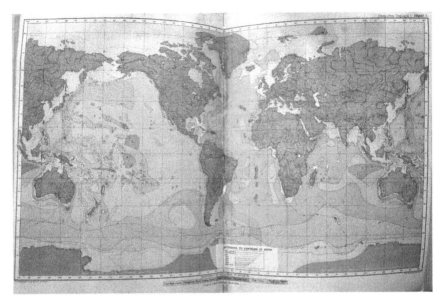

FIGURE 5.2. Deep-Sea Deposits. Chart 1. From John Murray and A.F. Renard, *Deep-Sea Deposits*, volume of *Report of the Scientific Results of the Voyage of H.M.S. Challenger during the Years 1873–1876* (London, 1891). NOAA Central Library Historical Collections.

Renard 1891, xxi). This alignment of depth, distance from shore, and nature of sedimentary fauna again allowed the naturalists to reexamine strata ashore. Since they now knew what conditions in the modern ocean were producing a layer of siliceous ooze, for instance, they could argue that the landscape that contained a siliceous stratum ashore must once have been under those same conditions. Examining the precise mix of fossilized fauna under a microscope could provide an even closer window onto the geological past, dictating an even narrower band of conditions in which the observed layers must have formed.

Over the three-and-a-half years of the *Challenger* voyage, the bottom became a familiar place to the naturalists. The various oozes and clays and their constituent parts entered the naturalists' everyday vocabulary. They were able to classify sedimentary layers as terrigenous (occurring in the littoral, in shallower water, and generally consisting of materials washed into the sea from the land), pelagic, or benthic, and they sketched them onto the contours emerging from the accompanying soundings. The result was a visualization of the bottom as a knowable, even colorful territory—with grey and blue and red clay, and chocolate oozes—delineated by depth and distance from shore (figure 5.2). The naturalists thus constructed the bottom as a site for study which they could visit virtually through the use of technology, though none of them could ever observe it in situ.

At the same time, their comparisons with shore-based paleontology helped them understand the inhabitants of the benthic landscape as multidimensional, existing in time as well as space, and in chemistry and geology as well as biology.

CLAIMING THE OCEAN BED

While the *Challenger's* primary mission during the expedition was scientific, it remained a naval vessel, so its naval officers were tasked with hydrographic studies of their own even as they worked to support the naturalists' efforts. The soundings and positional fixes that they obtained alongside and between each round of dredging and sampling allowed the naturalists to make exact claims about the provenance of their specimens and environmental data, but they also accumulated an expert knowledge about the ocean's dimensions, contributing to the long effort to map the globe.[27] The officers charted this positional data and returned it home to the Admiralty independent of the scientific results. Both these sets of knowledge—the positional fixes and the accumulated biological, geological, and environmental information associated with them—contributed to an ongoing British imperial project to "rule" the waves by thoroughly knowing them.

The British were not the only ones to stake these kinds of knowledge-based territorial claims. As new technologies were developed that allowed both navies and scientists to more accurately chart the bottom, they found each other to be good partners in the ongoing process of filling in the bathymetric map that Matthew Fontaine Maury had first begun. Along the way, they named the features they felt out along the bottom, and this process of labeling staked claims to a certain possession embodied in knowledge, even when attached to landscape features in a territory that could not be subject to literal territorial claims of ownership. The *Challenger* officers had named various bays and straits after their colleagues aboard—and the naturalists had reciprocated when assigning scientific names to the previously unknown fauna discovered in their dredge hauls—but the unveiling of new bottom topography provided whole new chains of mountains and valleys on which to impose the names of their discoverers. Challenger Deep in the Pacific was not the least of these, and the expeditions that followed in the next century continued in the same tradition.

In the second decade of the twentieth century, still smarting from their defeat in the Great War and from the punishing terms of the Treaty of Versailles that ended it, the German navy partnered with German oceanographers and the Notgemeinschaft der Deutschen Wissenschaft, a nongovernmental organization with a commitment to fund German science, to assemble a scientific expedition to study the Atlantic Ocean—the Deutsche Atlantische Expedition. The postwar context was important to their efforts, as both navy and scientists were suffering from real and perceived exclusion from what they saw as their rightful place among the superpowers. The Versailles terms forbade the navy to build or arm new warships. The treaty had also stripped Germany of its colonies, something the navy saw as symbols of great power status. The colonies had served another purpose for scientists, providing friendly and accommodating centers for field research. In the wake of the war, Germans felt—and to a degree were—locked out of international funding and research opportunities, so the loss of these colonial spaces for fieldwork and the loss of domestic funding brought on by the country's precarious postwar economic situation made scientists fear they would fall from the top tier of scientific research that Germany had occupied before the war (Crawford 1988; Kevles 1971). The oceans, however, provided a space for both doing science and demonstrating the German navy's continued ability to operate, and from which they could not be barred or restricted. These benefits proved sufficient motivation to scrape together the funding for a two-year expedition to thoroughly study the South Atlantic from the deck of the gunboat-turned-surveying-vessel *Meteor*.

The expedition was equipped to perform thorough oceanographic and meteorological work at three hundred sampling points, called stations, arranged along fourteen latitudinal cross-sections of the Atlantic, called "profiles," from about 20°N to below 60°S. At sea, a typical station involved the lowering of numerous thermometers and sample bottles via a large winch on the aft deck; another winch forward was available as a backup. Each carried 8,000 meters of aluminum-bronze stranded wire rope, with a high tensile strength and an 830-kilogram breaking strength. The aluminum-bronze alloy meant the line required no grease to prevent corrosion. Sample bottles were spaced along the line at intervals, their number determined by the depth, and each could be triggered in sequence by a falling weight to capture 1.25 liters of seawater. A four-liter bottle at the bottom was rigged to sample seawater with no contact with metal (Spiess 1985, 94–95). Laboratory facilities on board and later analysis ashore would analyze the seawater chemistry of these samples, including salinity. The lines carried reversing thermometers, both pressure-protected and unprotected, alongside the water samplers. Once the line reached the desired depth, it would be left for twenty minutes for the thermometers to register accurately, and a messenger weight sent down the line would trigger both the water sample capture and the thermometer's reversal, locking the temperature reading at depth. The resulting temperature and salinity data together provided useful information about the movement of water throughout the ocean. The comparison of the protected thermometer with the unprotected, which was thus affected by pressure, provided a calculated depth based on the known effect of pressure on temperature and the regular increase of pressure with depth, which could double-check the sounding line. The thermometers could be calibrated in port with a bucket of ice water. To check that the water samplers closed at the desired depth, the resulting samples were analyzed for their hydrogen ion concentration, as that too varies regularly with depth (Spiess 1985, 96, 99).

This was a more thorough study of the ocean's dynamical aspects than had been conducted before, but not otherwise completely novel—a difference in quality and quantity of data rather than kind. What was truly new on board the *Meteor* were two new echo-sounding devices: one of completely German origin, developed by the Kiel-based Signalgesellschaft and thus named the Signal sounder, while the other was a product of the Submarine Signal Corporation in Boston, Massachusetts, sold to the Atlaswerke in Bremen. The latter equipment was renamed the Atlas sounder, despite being of American origin and barely having even been tested by the Germans before its last-minute deployment (Höhler 2002a, 140). To back up and verify the state-of-the-art echo sounders, *Meteor* carried two line sounders: a Thomson sounding machine, originally developed by William Thomson, Lord Kelvin, in the 1870s, and a Lucas machine, designed in the 1880s.

Echo sounding had been proposed as a means of depth-finding quite early; by 1858 Matthew Fontaine Maury reported that a number of methods had been tried, including the use of explosives or bells to create the sound signal, but "out in 'blue water' every trial was only a failure repeated" (Maury 1856, 243–44). A device to find depth by using a sound pulse had been patented in Germany for navigational use as early as 1912, and in the wake of the *Titanic* disaster that April inventors and maritime officials experimented with the same concept to measure distances horizontally or find objects (such as icebergs) in the water surrounding a vessel.[28] The Great War's interference in civilian shipping retarded the spread of the navigational technology and obscured much of the other experimentation beneath the cloak of military secrecy. By the 1920s, though, the United States, France, and Britain all experienced some success in developing these tools and were deploying them in limited fashion aboard moving ships. In 1922, an American warship sounded a continuous profile across the Atlantic, resulting in a depth cross-section that displayed the ocean's bottom contours along one line, as if along a single latitudinal "slice." While this demonstrated that such a feat could be accomplished, it hardly cast much more light on the topography of the Atlantic basin than had Maury's few dozen scattered line soundings. The following year the U.S. hydrographic office published a bathymetric chart of the California coastline incorporating five thousand sonar measurements conducted by two warships over thirty-eight days, but even this, while it represented a vast increase both in area covered and in time and labor spent, barely touched the rim of the vast Pacific (Höhler 2002a, 136; "Echo Sounding" 1923; Schott 1923).

In addition to the Signal and Atlas sounders described above, *Meteor* carried two iterations of earlier acoustic sounders. The Behm sounder worked essentially along the principles Maury and others had suggested; a blast cartridge detonated at the surface, sending a pulse of sound into the water and starting a timer that the returning echo stopped. It was only useful to 750 meters and even then was not terribly accurate. The free sounder, also known as the bomb sounder, was another product of the Signalgesellschaft. It used the direct sound of an explosive, rather than its echo, to measure depths in fairly shallow water (less than 200 meters). An explosive designed to fall through water at a constant rate was released, and its descent timed until it exploded upon contact with the bottom. The duration of its descent allowed calculation of water depth, ignoring the speed of sound in such shallow depths (Hoheisel-Huxmann 2007, 58).

The two new, state-of-the-art sounders, though, promised much more precise measurement, and the Germans hoped to provide the first comprehensive survey of the Atlantic. The Signal sounder was successfully tested during a preliminary, trial expedition, but the Atlas was not ready in time and thus went directly into service during the main expedition. Both operated by the emission of a 1,050 Hz sound pulse, which reflected off the bottom and returned to the ship after a delay that depended on the depth beneath the ship's keel. A receiving membrane on the hull detected the reflected signal. The sounding apparatus calculated the depth automatically based on the delay, using a preset approximation of the speed of sound in seawater.[29] Soundings could thus be conducted in any depth, in all weather, while the ship maintained speed. At every oceanographic station, an old-fashioned line sounding verified the echo sounding results (Spiess 1985, 151).

As historian Sabine Höhler has pointed out, the use of a sound moving through water to calculate distance rendered the water itself the medium of measurement, instead of the lines and weights of the previous generations of sounders. The properties of seawater that affected the behavior of sound within it thus became important objects of study for the purposes of their participation in this technology, rather than just as an end in itself (Höhler 2002a, 122). This would prove immensely important to the future direction of physical oceanography as a field, turning the world's blue water navies into important patrons of oceanography and simultaneously guiding the direction of study as the field developed over the course of the next several decades.

The results were as good as could have been hoped. The two modern echo sounders performed flawlessly; they remained in continuous use during the thirty-month expedition with no technical difficulties, usually taking individual measurements every twenty minutes, which placed them at two- or three-mile intervals (Spiess 1985, 83). When the bottom topography seemed particularly interesting, the interval was shortened. The morphology thus charted often determined the location or interval of oceanographic stations, as it suggested the contours of ocean basins, information that could, when augmented with the thermal and chemical results gained by sampling, elucidate the movement of deep currents. The resulting charts of Atlantic topography represented a significant legacy of the expedition and formed the basis for a three-dimensional, bottom relief model of the South Atlantic displayed in the Berlin Museum of Oceanography (Höhler 2002b, 234–46).

As with most major expeditions, the voluminous results of the Deutsche Atlantische Expedition led to the publication of numerous volumes of results on the various subjects examined during the voyage. In them, German oceanographers used the sounding data to reimagine the ocean, in some cases assigning new names and features, in others rearranging the ocean bottom, naming new contours and assigning old ones to newly differentiated basins, as Theodor Stocks and Georg Wüst did in *Die Tiefenverhältnisse des offenen Atlantischen Ozeans*. New data led to a finer-grained geographical understanding, though Stocks and Wüst were careful to point out that it "must naturally remain hypothetical in some parts" (1935, 32). In places, these new features bore German names—as did the Alfred Merz Plateau, near Bouvet Island in the South Atlantic, named for the chief scientist and expedition leader who had died in the course of their voyage. Even when they did not, their publication by German scientists in charts of the ocean's basin labeled as products of German science constituted a symbolic claiming of territory, as Sabine Höhler has argued; if Germans could no longer claim colonies ashore, they could, in the act of mapping, claim the bottom of the ocean (Spiess 1985, 149; Höhler 2002b).[30] I have elsewhere extended her argument to suggest that the expedition's meteorological efforts similarly laid claim to the currents of the air over the Atlantic (Hardy 2017). While the meteorologists on board both wrote their own reports and sent data home for further analysis, much of their results volume consists of hundreds of pages of tabulated data, an assertion of German data dominance to support the symbolic seizure of aerial territory. German oceanographers used their technologically derived knowledge of the ocean bottom to assert their—and their nation's—continued membership in the top tier of science and thus their continued claim to be a Great Power.

CONCLUSION

Over the course of the nineteenth and twentieth centuries, scientists used increasingly sophisticated technologies to build a detailed picture of the ocean floor. In doing so, they turned that picture to the overlapping purposes of science, government, and industry, imagining it as, among other things, a cradle for telegraph cables, an embodiment of geological history, and a territory to be ordered and named. That is not to say that any of these actors believed their understanding of the ocean's contours to be final; Maury revised his mental image of the bottom quickly as new soundings undermined his previous conceptions, and the German oceanographers understood their maps, though based on the then-state-of-the-art technology and the most detailed yet produced, "must naturally remain hypothetical in some parts" (Stocks and Wüst 1935, 32). Yet at the same time, these scientists understood their maps to be true, and with them they fostered their nation's commercial destiny, found evidence for deep geological time, and defended national pride.

Our conception of the bottom of the ocean continues to say as much about the structures of human science and politics as it does about the ocean itself, as the fluid understanding of both three-dimensional ocean and the water that fills it are recruited to commercial, scientific, and thus political enterprises. Oceanographers have vastly improved understanding of the ocean, most notably with the mid-twentieth-century discovery that the Mid-Atlantic Ridge first charted on the *Meteor* is in fact part of a forty-nine-thousand-mile-long, global structure and the key to the modern conception of plate tectonics (Felt 2012, 251). Yet at the same time, oceanographers continue to point out the many ways in which the ocean bottom remains unknown. It is a trope of oceanography to note that we know more about the surface of the moon—or sometimes Mars or beyond—than we do about the bottom of the ocean. In the Cold War, this conception of the ocean as unexplored fueled pleas for funding in a field that saw itself competing with space exploration for support. Despite the various forms of public and private funding expended on efforts to know the ocean since, the bottom remains unknown enough to continue supporting arguments for further funding and ship time.

The water itself, too, retains a fluid identity: it is measured, and it is also a technology of measurement. Stefan Helmreich (2011), examining the popular modern conception of the ocean's depths embodied in the Google Ocean application, has noted the "odd sensory feature" experienced when a user "flies" under the virtual surface to view the bathymetry of the ocean floor: the water itself is not present. This is somewhat ironic, for Google's engine constructs its image of seafloor topography from mathematical data that is in turn derived from a combination of the measurement of minute variations in satellite-based radar measurements of the water's surface that reflect the topography below and echo-soundings of the bottom, as Helmreich notes (1226). This means that the medium of measurement is water, just as it was for the *Meteor*'s soundings in the 1920s, though Google's imagined ocean renders both water and math invisible.

In the twenty-first century, the global ocean and the water that fills it have assumed another role in the measurement and representation of the oceans among scientists who use their various properties—pH level, oxygen content, ice coverage—to index the changes associated with the warming global climate. Ocean acidification, oxygen depletion, and the changing albedo, shifting global currents, and rising sea levels produced by shrinking ice caps thus provide a new reimagining of the ocean as bellwether for changes that will affect the terrestrial portions of the globe in turn. They thus still fuel arguments over geological history, as many believe Earth has now entered a new epoch in which human activity is the major motive of change: the Anthropocene. In pointing to the scientific knowledge about the oceans while engaging the excitement of its remaining mysteries, scientists and the activists who rely on their work thus build arguments and appeals from ocean data, asking their audiences to consider the fluid identity not just of the watery two-thirds of the planet, but of the entire Earth.

REFERENCES

Adler, Antony. 2014. "The Ship as Laboratory: Making Space for Field Science at Sea." *Journal of the History of Biology* 47 (3): 333–62.

———. 2019. *Neptune's Laboratory: Fantasy, Fear, and Science at Sea.* Cambridge, MA: Harvard University Press.

Alaniz, Rodolfo John. 2014. "Dredging Evolutionary Theory: The Emergence of the Deep Sea as a Transatlantic Site for Evolution, 1853–1876." PhD diss., University of California, San Diego.

Andrewes, William J, ed. 1996. *The Quest for Longitude: The Proceedings of the Longitude Symposium, Harvard University, Cambridge, Massachusetts, November 4-6, 1993.* Cambridge, MA: Collection of Historical Scientific Instruments.

Brooke, George M. 1980. *John M. Brooke: Naval Scientist and Educator.* Charlottesville: University Press of Virginia.

Buchanan, J[ohn] Y[oung]. 1919. "A Retrospective of Oceanography in the Twenty Years before 1895." In *Accounts Rendered of Work Done and Things Seen.* Cambridge: Cambridge University Press.

Burnett, D. Graham. 2009. "Hydrographic Discipline among the Navigators: Charting an 'Empire of Commerce and Science' in the Nineteenth-Century Pacific." In *The Imperial Map: Cartography and the Mastery of Empire*, edited by James R. Akerman, 185–260. Chicago: University of Chicago Press.

Carpenter, William B. 1868. "Preliminary Report of Dredging Operations in the Seas to the North of the British Islands, Carried on in Her Majesty's Steam-Vessel 'Lightning,' by Dr. Carpenter and Dr. Wyville Thomson, Professor of Natural History at Queen's College, Belfast." *Proceedings of the Royal Society* 17: 168–200.

Caskie, Jaquelin Ambler, ed. 1928. *Life and Letters of Matthew Fontaine Maury.* Richmond: Richmond Press.

Crawford, Elisabeth. 1988. "Internationalism in Science as a Casualty of the First World War: Relations between German and Allied Scientists as Reflected in Nominations for the Nobel Prizes in Physics and Chemistry." *Social Science Information* 27 (2): 163–201.

"Echo Sounding: Test Carried Out by the U.S.S. 'Stewart' 20th to 29th June 1922." 1923. *Hydrographic Review* 1: 71–72.

Felt, Hali. 2012. *Soundings: The Story of the Remarkable Woman Who Mapped the Ocean Floor.* New York: Henry Holt.

Hardy, Penelope K., and Helen M. Rozwadowski. 2020. "Maury for Modern times: Navigating a Racist Legacy in Ocean Science." *Oceanography* 33, no. 3: 8–13. https://doi.org/10.5670/oceanog.2020.302.

Hardy, Penelope K. 2016. "Every Ship a Floating Observatory: Matthew Fontaine Maury and the Acquisition of Knowledge at Sea." In *Soundings and Crossings: Doing Science at Sea 1800–1970*, edited by Katharine Anderson and Helen M. Rozwadowski, 17–48. Sagamore Beach, MA: Science History Publications/Watson Publishing International.

———. 2017. "Meteorology as Nationalism on the German Atlantic Expedition, 1925–1927." *History of Meteorology* 8: 124–44.

Helmreich, Stefan. 2011. "From Spaceship Earth to Google Ocean: Planetary Icons, Indexes, and Infrastructures," *Social Research* 78 (4): 1211–42.

Hoheisel-Huxmann, Reinhard. 2007. *Die Deutsche Atlantische Expedition 1925–1927: Planung und Verlauf*. Hamburg: Convent Verlag.

Höhler, Sabine. 2002a. "Depth Records and Ocean Volumes: Ocean Profiling by Sounding Technology, 1850–1930." *History and Technology* 18 (2): 119–54.

———. 2002b. "Profilgewinn. Karten der Atlantischen Expedition (1925–1927) der Notgemeinschaft der Deutschen Wissenschaft." *NTM N.S.* 10: 234–46.

Hsu, Mei-Ling. 1988. "Chinese Marine Cartography: Sea Charts of Pre-Modern China," *Imago Mundi* 40: 96–112.

Kevles, Daniel. 1971. "'Into Hostile Political Camps': The Reorganisation of International Science in World War I." *Isis* 62: 47–60

Laloë, Anne-Flore. 2012. "Where is Bathybius Haeckelii? The Ship as a Scientific Instrument and a Space of Science." In *Re-inventing the Ship: Science, Technology and the Maritime World, 1800–1918*, edited by Don Leggett and Richard Dunn, 113–30. Burlington, VT: Ashgate.

Lewis, Martin W. 1999. "Dividing the Ocean Sea," *The Geographical Review* 89 (2): 188–214.

Lyell, Charles. 1831–33. *Principles of Geology*. 3 vols. London.

Maury, Matthew Fontaine. 1856. *The Physical Geography of the Sea*. 6th ed. New York: Harper & Brothers.

Mills, Eric. 2009. *The Fluid Envelope of Our Planet: How the Study of Ocean Currents Became a Science*. Toronto: University of Toronto Press.

Morrison, Dane. 2017. *True Yankees: The South Seas and the Discovery of American Identity*. Baltimore: Johns Hopkins University Press.

Murray, John, and A.F. Renard. 1891. *Deep-Sea Deposits*. Vol. of *Report of the Scientific Results of the Voyage of H.M.S. Challenger during the Years 1873–1876*, edited by C. Wyville Thomson and John Murray. London: printed by Eyre & Spottiswoode.

Portuondo, María M. 2009. *Secret Science: Spanish Cosmography and the New World*. Chicago: University of Chicago Press.

Rainger, Ronald. 2007. "Edward 'Iceberg' Smith and American Polar Oceanography." In *Extremes: Oceanography's Adventures at the Poles*, edited by Keith R. Benson and Helen M. Rozwadowski. Sagamore Beach, MA: Science History Publications.

Reidy, Michael. 2008. *Tides of History: Ocean Science and Her Majesty's Navy*. Chicago: University of Chicago Press.

Reidy, Michael S, Gary Kroll, and Erik M. Conway. 2007. *Exploration and Science: Social Impact and Interaction*. Santa Barbara: ABC-CLIO.

Rice, A.L. 1975. "The Oceanography of John Ross' Arctic Expedition of 1818: A Reappraisal." *Journal of the Society for the Bibliography of Natural History* 7 (3): 291–319.

Ross, Sydney. 1962. "Scientist: The Story of a Word." *Annals of Science*, 18 (2): 71–73.

Rouleau, Brian. 2014. *With Sails Whitening Every Sea: Mariners and the Making of an American Empire*. Ithaca, NY: Cornell University Press.

Rozwadowski, Helen M. 1996. "Small World: Forging a Scientific Maritime Culture for Oceanography." *Isis* 87 (3): 409–29.

———. 2005. *Fathoming the Ocean: The Discovery and Exploration of the Deep Sea*. Cambridge, MA: Belknap Press.

———. 2018. *Vast Expanses: The History of the Ocean*. London: Reaktion Books.

Schatzberg, Eric. 2018. *Technology: Critical History of a Concept*. Chicago: University of Chicago Press.

Schott, Gerhard. 1923. "Tiefseelotungen mittelst Echolot." *Annalen der Hydrographie und Maritimen Meteorologie: Organ des Hydrographischen Bureaus und der Deutschen Seewarte* 51 (August): 192–95.

Smith, Jason W. 2018. *To Master the Boundless Sea: The US Navy, the Marine Environment, and the Cartography of Empire.* Chapel Hill: University of North Carolina Press.

Spiess, Fritz. 1985. *The Meteor Expedition: Scientific Results of the German Atlantic Expedition, 1925–1927.* Translated and edited by William J. Emery. New Delhi: Amerind.

Sponsel, Alistair. 2016. "An Amphibious Being: How Maritime Surveying Reshaped Darwin's Approach to Natural History." *Isis* 107 (2): 254–81.

Stocks, Theodor, and Georg Wüst. 1935. *Die Tiefenverhältnisse des offenen atlantischen Ozeans: Begleitworte zur Übersichtskarte 1:20 Mill.* Berlin: Walter de Gruyter.

Thomson, Charles Wyville. 1878. *The Atlantic: A Preliminary Account of the General Results of the Exploring Voyage of the H.M.S. "Challenger" during the Year 1873 and the Early Part of the Year 1876,* vol. 1. New York: Harper & Brothers.

Thomson, Charles Wyville, William Benjamin Carpenter, and John Gwyn Jeffreys. 1873. *The Depths of the Sea: An Account of the General Results of the Dredging Cruises of H.M.SS. 'Porcupine' and 'Lightning' during the Summers of 1868, 1869, and 1870, under the Scientific Direction of Dr. Carpenter, F.R.S., J. Gwyn Jeffreys, F.R.S., and Dr. Wyville Thomson, F.R.S.* New York: Macmillan.

Williams, Frances Leigh. 1963. *Matthew Fontaine Maury, Scientist of the Sea.* New Brunswick, NJ: Rutgers University Press.

NOTES

1. While this chapter concentrates on Western efforts, other peoples had their own approaches to knowing the oceans. Pacific Islanders have a long tradition of navigation using methods that do not require graphic representation. Ming dynasty China had a large fleet that navigated over a significant portion of the global ocean in the medieval period, but this activity ended with the death of Emperor Zhu Di in 1424. Ironically, as Michael S. Reidy, Gary Kroll, and Erik M. Conway (2007, 4–5) have pointed out, this meant China turned away from ocean exploration just as the era of extensive European ocean navigation was beginning. Mei-Ling Hsu (1988) provides an analysis of the extant early Chinese navigational charts (96–112); however, historians have not yet produced a concentrated body of scholarship on the history of ocean mapping in Asia to the degree that has been done for the West. Jakobina K. Arch, personal communication to the author, May 29, 2019.

2. Historians and philosophers of technology have spilled much ink defining the term, but I use it here to mean tools, techniques, and systems employed to accomplish goals or embody knowledge. In this case, those goals might include measurement of certain data points, their employment to make arguments, or their communication in graphic form. As Eric Schatzberg (2018) and others have shown, this is a modern interpretation of the word, so I am not suggesting that it was a category that the historical actors considered here would have used to describe the things I call technology.

3. For more on the changing role of the ocean in the scientific, public, and political imagination, see Adler (2019).

4. It is important to note that Maury's scientific work was inextricably entangled with his lifelong support of white supremacy and the institution of racial slavery. See Hardy and Rozwadowski (2020).

5. The US Coast Survey should in theory have been printing charts, but it had experienced a series of slowdowns and setbacks since its founding in 1807 that prevented it from producing broadly useful charts. Also, its work was limited to the American coast and the Gulf Stream.

6. Maury to Captain F.H. Gregory, USN, USS *North Carolina*, New York, August 29, 1843, vol. 1, "Letters Sent" (LS), RG 78, National Archives Building (NAB), Washington, DC.

7. Deep-sea sounding efforts date back at least to Magellan, though he gave up without finding the bottom. John Ross sounded as deep as one thousand fathoms in Baffin's Bay in the late 1810s, and in 1840 his nephew James Ross sounded well over two thousand fathoms in the Southern Ocean. Rice (1975). Naturalist George Wallich sounded and dredged on an 1860 cable survey, but he complained about the lack of science being done on the trip (see Rozwadowski 2005, 93, 140–41).

8. Maury to Gregory, August 29, 1843, vol. 1, "Letters Sent," RG 78, NAB.

9. "Suggestions for the Attention for the Home Squadron," October 3, 1843, vol. 1, "Letters Sent," RG 78, NAB. The petition came from a committee formed at the 1844 annual meeting of the Association of American Geologists and Naturalists, of which Maury was a member; Maury, H[enry] D[arwin] Rogers, [Edward] Hitchcock, [James Pollard] Espy, [William Charles] Redfield, [James Dwight] Dana, and [Joseph P.] Couthouy to [John Y. Mason,] secretary of the navy, November 11, 1844, vol. 1, LS, RG 78, NAB.

10. Maury to Crane, October 25, 1845, vol. 2, LS, RG 78, NAB.

11. House of Representatives Report no. 449, March 15, 1842, 27th Congress, 2nd session; Maury to Warrington, Washington, DC, January 16, 1848, vol. 2, LS, RG 78, NAB.

12. Maury to Robert Walsh, US Consul, Paris, July 9, 1847, vol. 2, "Letters Sent," RG 78, NAB; Maury to Adams, November 17, 1847, vol. 2, LS, RG 78, NAB.

13. Maury to Warrington, September 1848, vol. 3, LS, RG 78, NAB.

14. Maury, *Explanations and Sailing Directions to Accompany the Wind and Current Charts*, 4th ed. (1852), 41–42, cited in Williams (1963, 180); Maury to Warrington, September 1848, vol. 3, LS, RG 78, NAB.

15. These chart series have been enumerated by the American Geographical Society Library at the University of Wisconsin Milwaukee, "Matthew Fontaine MAURY Ocean Charts at AGS Library," updated May 18, 2012, http://uwm.edu/libraries/wp-content/uploads/sites/59/2014/06/maury.pdf.

16. Brooke (1980, 55–56); Maury to Duncan N. Ingraham, chief of the Bureau of Ordnance and Hydrography, January 16, 1857, "Letters Received" (LR), RG 45, NAB.

17. Maury to J.C. Dobbin, Secretary of the Navy, February 22, 1854, in Caskie (1928, 110–112). As Maury was deeply religious, this was likely more than a turn of phrase; his science retained an element of natural theology that was already becoming old-fashioned in scientific circles, and a belief in American Manifest Destiny that was not. For the recent oceanic turn in historical understanding of Manifest Destiny, see, for example, Rouleau (2014); Morrison (2017); and Smith (2018).

18. Earlier expeditions had carried naturalists to sea, but their focus had largely been ashore, with the ship providing transportation to that terrestrial destination. However, Alistair Sponsel (2016) has recently argued that Charles Darwin's work on board HMS *Beagle* relied heavily on hydrographer's techniques.

19. In addition to the efforts of Maury's lieutenants and of John Ross and James Ross (see note 11 above), naturalist George Wallich sounded and dredged on an 1860 cable survey, but he complained about the lack of science being done. See Rozwadowski (2005, 93, 140–41).

20. For these other contributions, see, for instance, Laloë (2012); Mills (2009); Rozwadowski (1996); and Adler (2014).

21. A fathom is a measurement of depth equal to six feet.

22. While the sailors called them "scientifics" or philosophers, Thomson and company would have thought of themselves as naturalists (or chemists in the specific case) or natural philosophers (for all of them). The fields now broadly called science were beginning to acquire that title, but still fell under the rubric of natural philosophy in the nineteenth century, especially in Britain. While the term "scientist" had been coined jokingly in 1833 by William Whewell, it did not enjoy broad (and serious) use until the latter part of the century in the United States, and near or even after the turn of the twentieth century in Britain (where it was avoided in part because Americans used it). William Whewell, "Art.

III—On the Connexion of the Physical Sciences by Mrs. Somerville," *The Quarterly Review* 51 (1834): 58–61, cited in Ross (1962).

23. "Report of the Committee appointed at the Meeting of the Council held October 26th [1871], to consider the Scheme of a Scientific Circumnavigation Expedition," reproduced in Thomson (1878, 75).

24. S[amuel] P[hillips] Lee and H.C. Elliott, *Report and Charts of the Cruise of the U.S. Brig Dolphin, Made under Direction of the Navy Department* (Washington, DC, 1854), notes page at end; Bailey, "Microscopical Examination of Deep Soundings from the Atlantic Ocean," *Quarterly Journal of Microscopic Science* 3 (1855): 90. Quote is from the latter.

25. "Article XII: Recent Discoveries and Improvements in Science and the Arts," *The American Eclectic*, September 1841, 389.

26. By contrast, catastrophism posited a geological history divided into epochs by occasional global catastrophes, thus bringing about sharp and sudden changes.

27. In navigation, the latitude and longitude measured at a particular location at sea is used to fix a ship's position on a chart, so the combined data set for one location is called a fix.

28. Ronald Rainger (2007, 135–38) placed the beginnings of American interest in oceanography in this context. The U.S. Coast Guard established its International Ice Patrol in 1914 to monitor icebergs in the Arctic and North Atlantic, and a dynamical understanding of ocean currents was necessary to understand their movements.

29. This was 1470 meters per second for the Signal sounder, or 1490 m/s for the Atlas (Hoheisel-Huxmann 2007, 58).

30. In this article and elsewhere, Höhler's argument follows Bruno Latour's depiction of the collection of data into charts as a rendition of time and space into stable but portable form (Höhler 2002a).

Aquapelagic Malolos

*Island-Water Imaginaries
in Coastal Bulacan, Philippines*

Kale Bantigue Fajardo

A young Pinay (Filipina) wearing a pink short-sleeve t-shirt and grey shorts rides and maneuvers a raft, using a long wooden pole, on an estuary near Pamarawan Island in Malolos, Bulacan, Philippines. She is in the middle of the frame, and our eyes move towards the river slightly as the girl's weight tips the raft towards the right. The pole also directs our eyes towards the water because the girl is pushing off on the starboard side. The empty plastic bottles that resourcefully keep the raft afloat are also visible. This is a place where everything is used and nothing is wasted. The river water is a grey-army-green color on the day that she is photographed. JP Hernandez, a local Malolenyo who posts photographs on Instagram and who uses the name *tanogotchi* and the hashtag #Malolos, is the photographer of the Pinay girl rafting (figure 6.1). His photograph and others from his Instagram account significantly expand the visual archive of Malolos, a town-turned-city that has historically and primarily relied on *land-based* imaginaries to promote its importance in dominant Philippine nationalist historiography. Here, I am using Phillip Hayward's theorization of *aquapelagos* as an "assemblage of the marine and land spaces of a group of islands and their adjacent waters (Hayward 2012).[1] This aquapelagic focus on marine, island, and adjacent spaces is highly generative and useful and can be applied in similar locations and contexts where places are not usually associated with or known for marine cultural heritage or the hydrohumanities. Indeed, Malolos is *hegemonically associated* with its more *inland city center* where important colonial, postcolonial, and nationalist architectural structures are located. These structures were important to the Philippine anticolonial nationalist struggle for independence and sovereignty in the nineteenth century. This inland architectural bias can be seen, for example, in the city of Malolos's recent tourism campaign (inaugurated in 2015), which uses the slogan "Vamos a Malolos." In the

FIGURE 6.1. *Sail Away with Me* (Pinay girl rafting, Pamarawan, Malolos, Bulacan, Philippines). Photo used by permission © 2016 Jonathan Hernandez, @tanogotchi on Instagram.

first *o* in the word *Malolos*, the city government's graphic designer uses the arch or curve of a rising yellow sun shining behind the famous Barasoain Church. Barasoain Church was the site of an important political moment and act of resistance for Indigenous Indios, Indias, Tagalogs, and Mistisx in 1898. A new government and nation were born and emerged in Barasoain Church as the First Congress was convened for the new Republika ng Pilipinas, the Republic of the Philippines.

Precisely because Hernandez pays attention to Malolos's coasts, islands, seawater, aquamarine, and cross currents spaces, places, and mobilities (Fajardo 2011) — instead of simply focusing on the usual historical landmarks in the city center of Malolos, such as Barasoain Church—he significantly expands our understanding of what constitutes, animates, or enlivens Malolos. He frames and shoots an example of a contemporary Indigenous-precolonial- decolonial island-*and-water*—in other words, an *aquapelagic imaginary* of Malolos, Bulacan. This pivot and return towards aquapelagicity challenges the once-established and entrenched nationalist discourse in the Philippines that emerged from the dominant precolonial Philippine historiography. Hernandez's photograph is a perfect example of an aquapelagic imaginary, and can be read closely alongside other aquapelagic imaginaries from Malolos in the contexts of rapid urbanization, megaregionalization, industrialization, and the climate crisis. Hernandez and other like-minded cultural producers invite and urge Filipinx environmentalists, human rights advocates, artists, photographers, writers, academics, organizers, teachers, etc. to focus our attention *now* on Malolos's rivers, estuaries, coasts, islands, and seas, and on the humans and nonhumans there. Ultimately, the current emergencies resulting from massive urbanization, industrialization, and megaregionalization in Malolos during the climate crisis demand that local and diasporic Malolenyx and their friends and allies now advocate for, protect, and culturally preserve coastal Malolos and coastal Bulacan. Successful advocacy works through various media, such as art/documentary/ethnographic photography/ filmmaking and videography, as well as through literature, music, and other arts. A "salvage" anthropological and environmental justice orientation and practice —in the social sciences and the humanities—is now urgently needed, because coastal Malolos is vulnerable to outside national institutions, capitalist structures, and extreme weather and environmental events.

In what follows, I elaborate on how and why aquapelagic narratives and visual imaginaries of Malolos, which highlight rivers, the sea, and island ecologies, peoples, and economies, are significant, given the current situation of hegemonic nationalism, urbanization, and megaregionalization in Central Luzon, along with the present reality of local/global anthropogenic climate crises. Specifically, I am referring to and will address the massive and ongoing real estate and infrastruc- ture "development" projects, including a proposed new Manila international air- port in coastal Bulacan (to be discussed later on in this essay); rising seas and extreme storms and weather events that are a part of local/global climate emergen- cies; as well as the marine ecologies that are important to fisherfolks and the more- than-human: mangrove trees, seagrass, and fish, and how they are currently being threatened. In these contexts, aquapelagic imaginaries and narratives about Malo- los, which stress the marine ecologies of Malolos, are critical because they direct Malolenyx, other Filipinx, other local residents, and the various mix of people and stakeholders previously outlined to consider and re/connect Malolos to more regional Manila Bay environmental and climate justice concerns and movements. A persistent and creative re/focus on the aquapelagic ecosystems of Malolos also invites Malolenyx and Filipinx who celebrate Malolos to think and act beyond the Philippine nation-state. It encourages questioning of what constitutes national histories, landmarks, or cultural heritage, and how a narrow and limiting nation- alist orientation and cultural politics, with their subsequent nationalist optics, potentially and unproductively disconnect us from other forms of solidarities and social and environmental movements that stress contemporary *regional and global* concerns, and which desire a different kind of future for humans and the more- than-human (Marran 2017; Probyn 2016).

Some brief background here on my positionality and methodology: I was born in Malolos and immigrated to Portland/Gladstone, Oregon, with my fam- ily as a young child in the early 1970s. Throughout my life as a child, youth, and adult, I have been regularly traveling and returning to Malolos. Through these opportunities, I was privileged and honored to develop a translocal and transna- tional relationship to Malolos and also to nearby Metro Manila. In the preface to my book *Filipino Crosscurrents: Oceanographies of Seafaring, Masculinities, and Globalization*, I wrote about personal and familial connections to Malolos's riverine and sea locations (Fajardo 2011). From 2012 to 2015, I conducted ethnographic fieldwork in Malolos during my summer and winter breaks from the University of Minnesota, Twin Cities.[2] This research included investigating and learning about cultural heritage and architectural preservation in Malolos. Later, I expanded my research to explore the concept of "marine cultural heritage" by focusing on coastal Malolos sites such as Pamarawan (island), introduced earlier.

NATIONALIST IMAGINARIES OF MALOLOS

Malolos, Bulacan, Philippines, is a city situated northwest of Metro Manila. Although it is a city located on Manila Bay, Malolos is *not* usually associated with water—specifically, the bay, the sea, or rivers. The city is also not usually associated with rafts (or other types of boats) such as the raft in Hernandez's Instagram photograph (map 6.1). Most Filipinx associate Malolos with dominant Philippine revolutionary and nationalist histories because at the end of the nineteenth century, Malolos was the "birthplace of the Philippine nation."[3] It is a history repeated through formal education in classrooms across the Philippines. Many Filipinx travel to Malolos on pilgrimages or go on school field trips to honor past Philippine patriots and their nationalist struggles. In contrast, Malolos is not particularly popular with international tourists and travelers. They often prefer to drive though Malolos on their way to Baguio from Metro Manila. Or they simply bypass Central Luzon and go directly to the stunning beaches and islands of the Visayas or to gorgeous Palawan. Through a nationalist historical perspective, Malolos is normatively associated with the Indio (Native) and Mestizo (mixed race) *bayani* (patriots) who convened the Malolos Congress and declared independence from Spain in 1898. This radical political act established the first republic in Asia. The Malolos Congress was established at Barasoain Church.[4] During the Philippine anticolonial and revolutionary war against colonial Spain, Malolos served as the seat of power for the newly created Republika ng Pilipinas (the Republic of the Philippines).

As a Malolenyo/Filipino American living in the United States, when describing my birthplace and hometown in the Philippines, I (too) often evoke these nationalist narratives, and I usually explain (to non-Filipinx) that Malolos is similar to Philadelphia or Boston in the United States because all three historical cities have similar revolutionary and nationalist significances to their respective nations. This association between Malolos, historical architecture (the church), and the establishment of the Republic of the Philippines is not only memorialized in Malolos's *Vamos a Malolos* tourism campaign (discussed earlier), but also reinforced in Philippine currency. The old Philippine ten-peso bill features Barasoain Church

MAP 6.1. MAP of Bulacan Province showing the location
of the City of Malolos. Wikimedia Commons.

and the Blood Compact on one side, while two important revolutionary patriotic
figures and leaders, Andres Bonifacio and Apolinario Mabini, can be found on the
other side. This is significant because like other national currencies, paper currencies are reserved for national heroes, landmarks, and important historical events.

I (too) originally understood Malolos in this historically conventional nationalist manner; that is, ultimately and categorically, Malolos is the birthplace of the
Philippine Republic, with the correlate being that nothing else about Malolos
really matters. My mother was born and raised in Malolos and grew up in Barasoain, where the famous church stands. She grew up in the Chichioco-Conjuangco
ancestral *bahay na bato* (stone home), which is located on Paseo del Congreso
(Congress Street), a stone's throw away from the Barasoain Church. As a child
I was not yet aware of Malolos's place in Philippine history, nor was I aware of
the nationalist significance of Barasoain Church. I simply thought that Barasoain
Church was my mother and maternal grandmother's church, the church where
my mother attended mass every day as a child and teenaged girl because her home
was so close.

My father grew up in nearby Sto. Rosario, an adjacent *barangay* (village) across a
small river that encompasses the nearby Cathedral of the Immaculate Concepcion
and where the first president of the Philippines, Emilio Aguinaldo, maintained
his residency in 1898, resulting from the Philippine declaration of independence,
the defeat of the Spanish, and the newly established republic. In Spanish colonial
urban planning, the cathedral and connected plaza(s) are considered the heart
and center of Malolos because Catholicism was central to Spanish colonial governance (1521–1898). The cathedral's central positioning signifies its geographic and
political importance. During the Spanish colonial period, wealthy Chinese and
Mestizo families usually established their homes near the cathedral and the Malolos town center.

This Spanish colonial, now Philippine postcolonial, urban planning and geography facilitates a dominant nationalist optic of Malolos where the churches function as metonyms. Barasoain Church and the Malolos Cathedral are literally geographically central, and the churches symbolize Philippine nationalist and revolutionary histories and nation-state power, which emerged in Malolos. Contemporary Philippine state officials have continued to evoke Malolos's and Barasoain Church's importance. For example, during the 1998 Philippine Centennial, President Joseph Estrada was inaugurated at Barasoain Church, and on June 12, 2012, President Noynoy Aquino (whose family owns the Chichioco-Conjuangco ancestral home and also has roots in Malolos)[5] attended a "simple and solemn celebration of the 114th Independence Day" at Barasoain Church (Balabo 2012).

Teodoro Agoncillo's historical monograph *Malolos: The Crisis of the Republic* (1960) has also helped to cement Malolos's importance in Philippine historiography. Agoncillo is a founding historian of nationalist Philippine historiography, and his Malolos book significantly details important nationalist and revolutionary politics and related military events that helped to establish the new Philippine nation. Agoncillo writes about the then town of Malolos in this way:

> Symbolically, Malolos [was] chosen as the center of the [anti-colonial and nationalist] pattern, for in an age of crisis, it represented for the Filipinos not only their political ideals, but also their poetic dreams. Malolos [is] the symbol of sustained hope. Tagalog poetry has sanctified it as the citadel of freedom and has woven around it the legend of a people fighting for human dignity. (Agoncillo 1960, x)

Here, Agoncillo uses a metaphor of political solemnity and land-based solidity (i.e., a citadel, a fortress) to stress Malolos's foundational geographic and political positioning in the Philippine archipelago. Given the weight of his foundational nationalist historiography, Agoncillo's text discursively settles a very particular perspective on Malolos; that is, again, that it is only/primarily the birthplace of the republic. It is for this particular nationalist history of Malolos that in 2001, the National Historical Commission of the Philippines declared Malolos's town center, which revolves around the two cathedrals and also includes several *bahay na bato* (stone ancestral homes) in Kamistisuhan,[6] a national heritage district.

In 2011, a cultural heritage NGO called *Heritage Conservation Society* based in Manila threatened to remove Malolos from a list of important cultural heritage sites in the Philippines. The newspaper account discussed how the Malolos City government had inadequately preserved important local architecture, specifically, the many bahay na bato in the city, which represent an important Spanish-Philippine hybrid architectural style. The NGO argued that there was a lack of proper regulation regarding how to maintain dwellings in Kamistisuhan, located in barangay Sto. Niño, which is adjacent to the Minor Basilica of the Immaculate Concepcion. In Malolos (and other parts of the Philippines), bahay na bato represent an important national architectural style. Moreover, prominent local families usually own them and understand them to be their "ancestral homes" (Zialcita 1980).

AQUAPELAGIC IMAGINARIES OF MALOLOS

In contrast to Agoncillo's nationalist imaginary of Malolos, independent scholar Jaime Veneracion, in *Malolos: A Legacy of Its Past* (2010), provides an alternative local, autochthonous, precolonial, and decolonial water-based understanding of Malolos. Instead of overemphasizing the revolution, Veneracion engages an ecological approach to geography and history and suggests that, to understand Malolos, we need to better grasp the larger geography of Bulacan province. On the second page of his book, he includes a hand-drawn map of rivers and estuaries in Malolos. The veiny nature of the hand-drawn map is similar to the map (map 6.2) of the Pampanga River traversing the provinces of Pampanga, Bulacan, and Nueva Ecija in Central Luzaon. Veneracion elaborates on Malolos's geography, "There are three well defined areas in the provincial topography: uplands [mountains]; lowlands [part of the river delta] and littoral or shorelands . . . Malolos is a town of the delta and the littoral" (Veneracion, 2010, 11). In the seventeenth century, the rivers and bay were the main highways, and the modes of transportation were rafts and boats. For this reason, most of the town centers were located by great river systems. Rivers and place-names are significant to Tagalog people. In precolonial times, instead of colonial Spanish surnames that were assigned to Indios (Natives), locations were often used to identify people. *Tagalog* is an ethno-linguistic term that identifies the native people of central Luzon and their diasporas, as well as the language we speak. It is understood to be a combination of *taga ilog*, which means "from the river." In beginning with a more localized, indigenous, precolonial, and decolonial *ecological* approach (not a nationalist one), along with understanding Malolos in the context of Bulacan province and its relation to water, rivers, and coasts, Veneracion ultimately reveals an aquapelagic Malolos. This aquapelagic Malolos has been submerged by hegemonic nationalist discourse and imaginaries (Glissant 1999). In this aquapelagic Malolos, the island barangay (village) of Pamarawan (mentioned earlier) holds a prominent position. Veneracion explains:

> The prototype shoreline area is Pamarawan which proudly announces its presence to Manila Bay by its tall bell tower and a large settlement of boat people. It has its own talipapa [temporary fish market]. Its very name suggests something ancient. Pamarawan was derived from the pre-Spanish Malayo-Polynesian term for boat (Parau or prahu). It means the "landing for boats," as in fact is a feature of the place up to now. Of the settlements of the shoreline, it has the largest conglomeration of bangkas [canoes]. (Veneracion 2010, 11)

MAP 6.2. DRAINAGE Map of the Pampanga River. The
Pamapanga River traverses the provinces of Pampanga,
Bulacan, and Nueva Ecija. Map by Felipe Aira, Wikimedia
Commons.

Like rivers, boats and canoes—linguistically and culturally—are highly significant
for Tagalogs to understanding local Philippine social relations (Fajardo 2011).
Barangay means village, and it is derived from the word *balanghai*, the word
for the ancient indigenous boats that allowed clans to travel. Boats signify this
indigenous notion of collectivity and community that forms on both water
and land.

Veneracion further reveals an Indigenous, precolonial, and decolonial ecological and social perspective on Pamarawan, and ultimately of Malolos (because
Pamarawan is part of Malolos). Subsequently, he invites us to see and appreciate Pamarawan and other river-and-coastal barangay, which are oriented toward
Manila Bay or the sea and *not* Malolos's Spanish- and Mestizx-influenced town
center. Veneracion's focus on water ecologies and his alternative local, native,
decolonial optics or ways of seeing, imagining, and historicizing, suggest to us
a complex aquapelagic island-water-based understanding of what can constitute
"cultural heritage" and the importance of water ecologies.

When I conducted ethnographic fieldwork in Malolos, I began to see how
local Malolenyx shared Veneracion's ecological awareness. My first example from
fieldwork, which suggests a twenty-first-century aquapelagic imaginary of Malolos, comes from the magazine *Biyaheng Malolos* (*Travel Malolos* 2010). I purchased a copy of *Biyaheng Malolos* from an independent t-shirt store in Malolos's
town center near the Minor Basilica of the Immaculate Concepcion in barangay
Sto. Niño. Imagine a small t-shirt shop filled with different-colored, limited-run
t-shirts— not of Ché Gueverra, Frida Kahlo, or Bruce Lee, but of Bulacan revolutionary figure and journalist Marcelo H. del Pilar. Imagine, too, brightly colored
t-shirts that say "Bulakenyo" or "Bulakenya," or t-shirts that say "Bulacan State
University" on them. I bought a "Bulakenyo" t-shirt for myself and one for my

proud Bulakenyo father (who along with my mother retired part-time in Malolos [but she probably would not have wanted a t-shirt.]) As I was paying, I noticed a stack of slick glossy magazines on the counter and began to thumb through one. I again bought two copies (one for me, one for my dad). Later, I learned that the young Filipino owner of the shop helped to produce the magazine.

The cover depicts a local Malolos jeepney, a working-class street-level example of Philippine design and popular culture. Jeepneys are an affordable form of local transportation that can be found in Malolos and other parts of the Philippines. They are associated with the *masa* (working class and poor masses in the Philippines). Barasoain Church is on the cover, but it is located in the background, significantly smaller than the jeepney. By receding the historical church, the magazine cover suggests a generational difference in geographic orientation and imagination. That is, I hypothesize that the editors, writers, and photographers of the magazine suggest that their cultural productions will not be traditional like their elders' Malolos, where national patriotism is the primary cultural and historical orientation and commitment. In the upper-left corner, there is a description of the contents/theme of the magazine: "fashion, culture, heritage." *Biyaheng Malolos* is an example of what might also be called "Millennial Malolos." The glossy magazine, which is similar to magazines coming out of Manila and other global cities (and which is something unprecedented for Malolos), was being sold inside an independent t-shirt shop that "represents" Malolos. Here, I use *represent* in the hip-hop sense of the word. The shop and magazine gesture towards a connection to broader local, regional, and/or global youth cultures, which emphasize taking pride in the local. The magazine includes layouts of models wearing t-shirts that say "Probinsiyano" or "Probinsiyana," which conveys pride in being "from the provinces," not from the megalopolis of Manila. Historically, the provincial spaces have been constructed as backward/rural spaces compared to (more) "cosmopolitan Manila." Inside *Biyang Malolos*, the editors and writers feature Barasoain Church, the Kamistisuhan and the heritage district there, but significantly, they go beyond the town center and include coastal Malolos. They also feature Pamarawan (Island)—the landing place of boats, rafts, and canoes.

Written by Josefina Alcaraz, photographs by Cenen Pangilinan and Ian Cruz, design layout by Cenen Pangilinan, the four-page spread that begins on page 56 of *Biyang Malolos* is titled "Unravel the Isle of Pamarawan." The opening photograph extends across the first two pages of the piece and includes a panoramic photograph of indigenous Tagalog *bangka* (canoes), just as Veneracion describes, and they are lined up on the shore with the Port of Pamarawan behind the canoes. To the left are the waters of Manila Bay, to the right, the estuary that connects Pamarawan to Atlag, barangay Panasahan, and inland Malolos. The next two pages include photographs and text that depict daily life in Pamarawan. Alcaraz, Pangilinan, and Cruz also discuss other areas of island locations in coastal Malolos: Namayan, Babatnin, Masile, and Caliligawan.

Alcaraz writes: "Carefully-lined salted and dried fish, parked motored boats, natives having chit-chats with neighbors, children freely running around, a pack of women doing *hayuma*—an indigenous way of stitching nets to produce a larger one, fishermen pulling up their fish nets for the day's catch, are very typical scenarios that would best describe Pamarawan's community" (Alcaraz 2010, 56–59). Here, Alcaraz and the editors of *Biyaheng Malolos* emphasize an aquapelagic Malolos, not the hegemonic nationalist imaginary previously described. Fish, boats, fisherpeople, and everyday indigenous (not Spanish) marine/aquatic practices are highlighted, inviting readers, visitors, and travelers (*Biyang Malolos* is a travel magazine, after all) to expand their understanding of Malolos's heterogenous cultural heritages and to appreciate Malolos's aquapelagic spaces and places.

Biyaheng Malolos's attention to Pamarawan and women engaged in *hayuma* (the local practice of stitching nets to produce a larger one) is connected to Hernandez's photograph, as the rafter is also female. Filipino boys and men usually dominate boats, canoes, and industrial container ships in Philippine maritime representations (Fajardo 2011). The Pinay rafter moves her craft constructed of recycled plastic, signaling thriftiness and ecological awareness through the reuse of the plastic bottles. In a conversation with Hernandez, I asked him about his Pamarawan-based photography, including the Pinay rafter. He explained:

> I took the shots during one of our tours to Pamarawan. [My wife] Rheeza is part of a tour operator business and they take tourists to Pamarawan to experience coastal life—what people do there for a living, the food there and how they cope with living on an island. I took pictures as we went along. *Taking note of things you don't necessarily see on the mainland.* I wanted to convey in my photographs that this is what you'll see if you go to Pamarawan. Photographing Malolos makes me tell the audience that this is the past, especially the heritage pictures, and I also photograph for posterity. And for others to be aware, so that they will take care of our heritage. (Emphasis added)

Hernandez's informal artist and mission statement reveals his ecological awareness of Malolos as a city of the river delta, shorelines, and bay, as well as his awareness that aquapelagic life in Pamarawan engages all genders, not just men and boys. He and his wife Rheeza seek to promote small-scale tourism to coastal barangays, where marine culture, fisheries, and Malolenyx cuisine are experienced, usually by visitors from Metro Manila. This marine-based tourist experience is a more recent development, something that did not occur during my childhood, youth, or earlier periods of adulthood when I traveled to Malolos regularly, or even when I first began my fieldwork in Malolos. In his explanation, Hernandez also underscores the necessity to photograph Pamarawan and other coastal and river barangays. He uses the word *posterity*, thus evoking future generations who need to know and remember Pamarawan and other coastal Malolos ecologies. Equally important, he seeks to develop an ethos of care in *all* who see and look at his photos.

In another photograph, Hernandez shoots a dazzling sunset from a bridge looking down on a river in Atlag, a riverine/estuary-based barangay, where waters connect to Pamarawan. The sunset includes layers of oranges, with the deepest at the horizon, with hazy blue skies and clouds near the top. Hernandez, positioned higher up on a bridge, looks down, on to several large and long canoes that transport people between coastal and insular barangays like Pamarawan and mainland Malolos. Hernandez's photograph recalls and dialogues with my earlier autoethnographic imaginary from "Boyhood and Boatmen," the preface to *Filipino Crosscurrents* (Fajardo 2011, ix–xiii). In future discussions of imaginaries of Malolos, as a Malolenyo living in the United States and working in solidarity with Malolenyx living in Malolos, I would be honored if some of my work is included in the new Aquapelagic Malolos archive that I have assembled and have been commenting upon, and which I hope other scholars, researchers, artists, and writers will continue to develop further. In "Boyhood and Boatmen," I dialogue and evoke Pablo Neruda's poem "The First Sea," which describes the poet's first encounter with the sea as a boy (Neruda 2004). Inspired by Neruda, I recall an important childhood memory of first encountering the sea and maritime Filipino masculinities in Atlag. Prior to my immigrating to the United States and Portland, Oregon, as a child, my paternal grandfather took me with him to Atlag to visit relatives. (My father's side of the family has roots and routes in Sto. Rosario *and* Atlag.) I later describe how he took me to the outdoor "wet market" in Atlag and showed me the fish and other seafood being sold there. I speculate that my grandfather did this because he was a proud fisherman.

THREATS TO MALOLOS: URBANIZATION AND MEGAREGIONALIZATION

In an ethnography titled *Neoliberalizing Spaces in the Philippines: Suburbanization, Transnational Migration, and Dispossession* (2016), Arnisson Andre Ortega describes the "urbanization revolution" that is happening in the southern part of Manila, now technically categorized as a megacity. (Megacities are cities with populations of ten million or more.) Ortega uses the term *megalopolis* to evoke cities such as Metro Manila, which at the time of Ortega's book's publication (2016) was the fourth most populous megacity in the world.[7] Of the world's megacities, Metro Manila is the densest. Metro Manila's population is approximately thirteen million, but in different accounts Metro Manila's population is sometimes noted as high as twenty-three million. The numbers appear to depend on how demographers understand the megacity's boundaries. "Mega Manila" is sometimes used to refer to Manila and its relationship or encroachment into nearby provinces, so it is more of a transregional designation.

Ortega documents and analyzes the intensified real estate development boom and related population increases in the southern part of the megalopolis where he is from. In my opinion and based on observations in Malolos and Central Luzon, a similar situation is occurring north of Manila, perhaps even more so, as the Philippine state tries to connect Metro Manila to shipping and naval facilities in Olongopo City and Subic Bay, as well as to cities in the North such as Angeles, where Clark International Airport is located. The state and capital attempts to connect these areas in Central Luzon, in order to develop a megaregion. Ortega emphasizes the rise of gated communities, which he sees as a marker of urbanization, resulting from increased wealth, the presence of Overseas Filipino/as, and transnational capital. In addition to gated communities, the development of malls must also be considered, as they are also markers of urbanization in the Philippines, especially since Manila is considered the "Mall Capital of Asia" (Yao 2010). During fieldwork in Malolos, two Manila-style malls opened—*The Cabanas* and *Robinsons Place Malolos*. Prior to these commercial developments, Malolenyx had to travel about one hour to Quezon City in Metro Manila to go to the nearest large mall. This suggests how Metro Manila continues to creep out into other provinces, expanding into towns and rural areas. In Malolos, gated communities and commercial developments have been built in former rice paddies and fish ponds, which were once green spaces that nourished humans and nonhumans, such as birds and fish.

In addition to the "loss, fragmentation, and the degradation of habitats" (PEM-SEA 2012), increased population in the Manila Bay region also creates "overexploitation of resources for livelihoods and commercial purposes" and contributes to more pollution (e.g., "chemical fertilizers and discharge of domestic sewage, but also from toxin such as pesticides and hazardous chemicals"). The national fisherfolk alliance Pambansang Lakas ng Kilusang Mamamalakaya ng Pilipinas (Pamalakaya-Pilipinas), for example, cites a 2015 Ocean Conservancy study that reveals that "74% of plastics that ended up in the sea came from previously collected garbage" (Salamat 2019). Pamalakaya-Pilipinas also cites the "failure of [the Philippine] government . . . to provide . . . significant water and solid waste treatment and management," which contributes to the degradation of communities and habitats.

Rapid urbanization and megaregionalization continue to expand and intensify in Malolos and in other parts of Bulacan. Previous Bulacan governor Wilhelmino Sy-Alvarado has officially stated and proposed that Bulacan follow the Shenzhen model (initiated by late People's Republic of China leader Deng Xioping). The governor stated in 2013 that his plan has not been finalized but that he is "looking forward to replicating its industrialization and modernization experience" (Balabo 2013). In 1979, Shenzen was a small fishing village with fertile agrarian land with a population of thirty thousand. As of 2012, Shenzhen, located north of Hong Kong, was a sprawling and high-tech manufacturing and service city. As I finalize this essay, signs of rapid urbanization in Bulacan province continue in disturbing ways. In late 2019, the San Miguel Corporation was scheduled to break ground on the new Manila International Airport in Bulakan, Bulacan, which is adjacent to Malolos. This airport complex, also referred to as the Aerotropolis, is scheduled to be built in barangay Taliptip, another Manila Bay and Bulacan coastal area, which is adjacent to Pamarawan. The Aerotropolis is supposed to replace highly congested Ninoy Aquino International Airport in Pasay, Metro Manila. When completed, the new airport is supposed to serve one hundred million passengers each year (Macaraeg 2019). The first phase of construction began in October 2020 and is scheduled to open in 2026 (Gonzales 2020).

THE LOCAL/GLOBAL CLIMATE CRISIS

The Aerotropolis infrastructure development project just introduced is being proposed and planned in the context of the local/global anthropogenic climate crisis. The massive infrastructure project seriously threatens the human and more-than-human communities of coastal Malolos, coastal Bulacan, and broader Manila Bay. It is scheduled to be built on a twenty-five-hundred-hectare area in Taliptip, Bulakan, along Manila Bay, extremely short-sighted planning: the Philippines has already experienced extreme weather, such as Super Typhoon Haiyan/Yolanda in 2013, which destroyed Tacloban, while Typhoon Ketsana submerged Manila in 2009, and in 2011, Typhoon Nesat created huge storm surges in Manila Bay (Robles 2013). In 2012, there was also a historic habagat (the humid southwest monsoon winds that regularly blow from June through July), which brought severe rains and flooded Malolos's coastal barangays, as well as more city center villages such as Sto. Rosario and Barasoain, discussed earlier.[8]

Pamalakaya-Pilipinas strongly opposes the Aerotropolis project. Indeed, the activist fisherfolk group calls it "the biggest environmental disaster to hit Manila Bay." Pamalakaya-Pilipinas Chairperson Fernando Hicap explains, "This project will not only destroy marine ecosystem, but also the livelihood of thousands of fisherfolk who subsist by fishing in the rich fishing hub of Manila Bay." Seven hundred fishing and coastal families in Taliptip will be displaced. The area is also home to "22 types of mangrove, including *Piapi*, a firm type of mangrove that serves as a natural wave barrier and shelter for fish" (Hicap 2018). When approximately six hundred mangrove trees were cut in Taliptip in 2018, local residents reported that they had "suffered unprecedented flooding and soil erosion" (Salamat 2019). Mangroves protect coastal areas during typhoon season. Without this natural barrier, there are stronger storm surges, increased flooding, and loss of habitat for fish and other animals. The project will also negatively impact salt-making livelihoods that are important to coastal villages in Bulacan.

Hicap explains the role of Philippine president Rodrigo Duterte in the proposed environmental disaster: "President Duterte is on a roll in selling our coastal waters to big-time business magnates; not less than 5 destructive reclamation projects have been approved in Manila Bay since he assumed post as chief executive, all at the expense of socio-economic rights of hundreds of thousands of fisherfolk and coastal settlers" (Pamalakaya-Pilipinas 2018).

CONCLUSION: WATER CONNECTS

While nationalist historiography narrows perspectives on Malolos, emphasizing land, the city center, the built environment, postcolonial architecture, and past revolutionary patriots, my essay seeks to counterbalance nationalist historiographies and imaginaries of Malolos by stressing alternatives—specifically, more aquapelagic and marine-based perspectives, narratives, visual imaginaries, and knowledge. The collective efforts of Hernandez, Veneracion, Alcaraz, Pangilinan, Cruz, Fajardo, and Pamalakaya-Pilipinas stress an aquapelagic Malolos and an aquapelagic Bulacan province, which help to document, bear witness to, and raise awareness about what is being increasingly destroyed or flooded in coastal Malolos and Bulacan areas, and in the larger Manila Bay region in the context of rapid urbanization, megaregionalization, industrialization, and the local/global anthropogenic climate crisis. Indeed, the proposed Aerotropolis project in Bulakan, Bulacan, makes these cultural production and activist efforts even more meaningful, timely, and urgent.

The aforementioned cultural producers and activists remind the Philippine nation, its citizens, and all who visit Malolos and Bulacan that there is more to Malolos than the Spanish colonial churches and colonial/postcolonial stone homes located inland, in mainland Malolos. Their/our cultural productions and activist discourse may not be circulating in large numbers (yet), nor are our visual imaginaries established or hegemonic. However, the aquapelagic Malolos archive that I have assembled and analyzed here reminds us that there are important marine-based coastal ecosystems and communities in Malolos and broader Bulacan that need documentation, advocacy, preservation, and protection. Historic Barasoain Church and other inland architecture where native and local revolutionaries politically decolonized the Philippines and established the first republic in Asia in 1898 are certainly important cultural heritage landmarks, and these structures and histories are important to the city of Malolos and for Malolenyx cultural identities. However, it is imperative that we also engage with and protect lesser-known Pamarawan and other coastal areas, which are geographically and culturally oriented toward the sea, to Manila Bay, and beyond. In this volume, Penelope K. Hardy reminds us that "humans have always interacted with the sea, [but] the ways in which they have imagined and thus defined it have changed significantly over the course of history." This essay tries to account for how the sea and coastal Malolos and coastal Bulacan ecologies are being newly created, imagined, documented, and assembled. The aquapelagic Malolos tropes used in contemporary historical accounts, ethnography, travel writing, photography, and in activism represent a significant historical, cultural, and political shift in how Malolos and its waters and islands are being imagined in the Philippines and in the diaspora, especially as increased flooding, rising seas, and more extreme weather are the new normal in the Philippines.

In a 2014 report titled "The City of Malolos: Towards a Local Climate Change Action Plan," researchers Zoë Greig, Leanna Leib-Milburn, Victor D. Ngo, and Meika S. Taylor state that the challenges Malolos faces include "increased severity and frequency of flooding; water scarcity; sea level rise; salt water intrusion; land subsidence; and low ecosystem and watershed health" (Greig et al, 2014, 46). Aquapelagic narratives about Malolos and marine-based photography of coastal Malolos locations, which stress the river, sea, and island ecologies of Malolos, are important because rapid urbanization, megaregionalization, industrialization, and the local/global anthropogenic climate crisis seriously threaten coastal communities—comprised of humans and the more-than-human—and biologically diverse ecosystems. Aquapelagic imaginaries of Malolos direct Malolenyx, other Filipinx, and other local residents and allies to connect Malolos to other environmental justice struggles in places such as Taliptip in nearby Bulakan, Bulacan, where the destructive Aerotropolis transportation infrastructure is being planned. Just as importantly, a focus on aquapelagic island-water ecologies also remind Malolenyx and other Filipinx who love and celebrate Malolos to think and act *beyond* the Philippine nation-state. Aquapelagic Malolos narratives and visual imaginaries in the context of local/global anthropogenic climate change can potentially mobilize Malolenyx at home and abroad to consider and act in solidarity with other local and national environmental and climate justice movements, but they also compel and direct us to actively participate in global environmental justice and climate justice movements. Attention and care for Malolos's marine cultural heritage and ecologies also direct us to consider and act in solidarity with humans *and* the more-than-human—that is, oceans, coastlines, mangrove trees, seagrasses, birds, and fish (Marran 2017; Probyn 2016).[9] While the building of the Aerotropolis offers and paves the way for a particular kind of globality and mobility (produced via air travel), local-global movements for environmental and climate justice offer an alternative. Defiantly, fisherfolk advocate and leader Hicap reminds us what is at stake if urbanization and megaregionalization continue in coastal Bulacan without care, protection, and resistance: "24 hectares of fishing reservation areas will be wiped out because of the land reclamation, this also mean [sic] loss of the traditional fish species and fish catch of small fisherfolk in the province of Bulacan." Pamalakay-Pilpinas "opposes this grand sellout of our traditional fishing waters at all cost. None of the fisherfolk deserves to be ejected from their communities just to pave way for an international airport and metropolis that will only benefit a few developers" (Hicap 2018). In shifting our attention, advocacy, and activism towards coastal Malolos and coastal Bulacan and paying closer attention to the aquapelagic, we are better able to see and imagine a more complex Malolos and Bulacan province, one that emphasizes islands, coasts, estuaries, rivers, and Manila Bay and the less resourced fisherpeople, as well as other local residents who are sustained by the sea. Equally important, a focus on an aquapelagic Malolos and Bulacan Province reminds us to also consider, take care of, and live well with the fish, birds, and mangroves that are equally important in and to Malolos and Bulacan ecologies, as well as the ecologies of the broader Manila Bay region.

ACKNOWLEDGMENTS

I want to thank my colleagues at the University of California, Merced, especially Ma Vang, Christina Lux, Ignacio López-Calvo, Kim De Wolff, and Rina C. Faletti for inviting me to give a lecture on my Malolos research at the Center for the Humanities Water Seminar in 2017 and which I eventually revised as this essay. Faletti, De Wolff, and the anonymous reviewers provided me with important editorial suggestions, which greatly improved this essay. Thank you to all of the editors and contributors to this volume, as well as the editors and staff at the University of California Press who all made this book possible.

REFERENCES

Agoncillo, Teodoro A. 1960. *Malolos: The Crisis of the Republic*. Diliman, Quezon City: University of the Philippines Press.

Alcaraz, Josefina. 2010. "Unravel the Isle of Pamarawan," *Biyeng Malolos* 1 (1): 56–59.

Balabo, Dino. 2012. "P-Noy to Lead Simple Independence Day Rites." *Phil Star Global*, June 10, 2012. www.philstar.com/headlines/2012/06/10/815544/p-noy-lead-simple-indepen dence-day-rites-barasoain-church.

———. 2013. "Bulacan to Copy Shenzen Model for Development." *Punto.com.ph*, February 21, 2013. https://punto.com.ph/bulacan-to-copy-shenzhen-model-for-development/.

Boellstorff, Tom. 2006. *The Gay Archipelago: Sexuality and Nation in Indonesia*. Princeton: Princeton University Press.

Fajardo, Kale Bantigue. 2011. *Filipino Crosscurrents: Oceanographies of Seafaring, Masculinities, and Globalization*. Minneapolis: University of Minnesota Press.

Glissant, Édouard. 1999. *Caribbean Discourse: Selected Essays*. Charlottesville: University of Virginia Press, 1999.

Gonzales, Iris. 2020. "SMC to Start Construction of Bulacan Airport." *Philippine Star*, October 7, 2020.

Greig, Zoë, Leanna Leib-Milburn, Victor D. Ngo, and Meika S Taylor. 2014. "The City of Malolos: Towards a Local Climate Change Action Plan." https://urbanizingwatersheds .files.wordpress.com/2015/05/city-of-malolos_towards-a-lccap_2014.pdf.

Hau'ofa, Epeli. 1993. "Our Sea of Islands." In *A New Oceania: Rediscovering Our Sea of Islands*, edited by Vijay Naidu, Eric Waddell, and Epeli Hau'ofa. Suva: School of Social and Economic Development, The University of the South Pacific.

Hayward, Phillip. 2012. "Aquapelagos and Aquapelagic Assemblages." *Shima: The International Journal of Research into Island Cultures* 6 (1): 1–11.

Hicap, Fernando. 2018. "'Bulacan Aerotropolis' Threatens Marine Ecology, Food Security–Fisherfolk" (press release). *Stay Grounded*, September 30, 2018. https://stay-grounded .org/press-release-bulacan-aerotropolis-threatens-marine-ecology-food-security -fisherfolk/.

Joseph, May. 2013. *Fluid New York: Cosmopolitan Urbanism and the Green Imagination.* Durham, NC: Duke University Press.

Macaraeg, Pauline. 2019. "San Miguel to Start Building New International Airport in Bulacan This Year." *Esquire*, January 8, 2019. www.esquiremag.ph/money/san-miguel-to-build -bulacan-international-airport-this-year-a00287–20190108.

Marran, Christine. 2017. *Ecology without Culture: Aesthetics for a Toxic World*. Minneapolis: University of Minnesota Press.

Neruda, Pablo. 2004. *On the Blue Shore of Silence*. New York: Rayo Press.

Pamalakaya-Pilipinas. 2018. "Fisherfolk Say No to Bulacan International Airport." *Pamalakaya* (blog), April 4, 2018. https://pamalakayaweb.wordpress.com/2018/04/04/fisherfolk -says-no-to-bulacan-international-airport/.

PEMSEA (Partnerships in Environmental Management for the Seas of East Asia). 2012. "Integrating Climate Change and Disaster Risk Scenarios into Coastal Land and Sea Use Planning in Manila Bay." http://pemsea.org/publications/reports/integrating-climate -change-and-disaster-risk-scenarios-coastal-land-and-sea-use.

Probyn, Elspeth. 2016. *Eating the Ocean*. Durham, NC: Duke University Press.

Roberts, Brian, and Michelle Ann Stephens. 2017. *Archipelagic American Studies*. Durham, NC: Duke University Press.

Robles, Alan. 2013. "If a Super Typhoon Hit Manila, Damage Would Be as Bad or Worse." *South China Morning Post*, November 13, 2013. www.scmp.com/news/asia/article /1355407/if-super-typhoon-hit-manila-damage-would-be-bad-or-worse.

Salamat, Marya. 2019. "Bulacan Fisherfolk, Women Want Genuine Inclusive Manila Bay Rehabilitation." *Bulatlat*, February 27, 2019. www.bulatlat.com/2019/02/27/bulacan-fisher folk-women-want-genuine-inclusive-manila-bay-rehabilitation/.

Tiongson, Nicanor. 2004. *The Women of Malolos*. Quezon City: Ateneo University Press.

Veneracion, Jaime. 2010. *Malolos: A Legacy of Its Past*. Baliuag: MSV Printers & Publishers.

Yao, Deborah. 2010. "Shopping in Manila, the Mall Capitol of Asia." *Seattle Times*, April 11, 2010. www.seattletimes.com/life/travel/shopping-in-manila-the-mall-capital-of-asia/.

Zialcita, Fernando N., and Martin Tinio, Jr. 1980. *Philippine Ancestral Houses (1810–1930)*. Oakland, CA: Masala Press.

NOTES

1. Hayward's notion of the aquapelagic dialogues well with Epeli Hau'ofa's notion of the Pacific Basin and Region as a "Sea of Islands" (1993); Tom Boellstorff *The Gay Archipelago* (2006); May Joseph's *Fluid New York* (2013); the *Archipelagic American Studies* anthology (2017), edited by Brian Roberts and Michelle Ann Stephens; as well as my book *Filipino Crosscurrents: Oceanographies of Seafaring, Masculinities and Globalization* (2011) (to give a few important examples of how scholars are thinking through islands, water, and archipelagos). A special thank you to May Joseph, who told me about Hayward's generative essay.

2. My fieldwork was generously funded by a University of Minnesota Grant-in-Aid of Research, Artistry and Scholarship.

3. *Filipinx* is a gender inclusive and non-gender-binary term; it replaces the binary term *Filipino/as.*

4. The church is officially named Our Lady of Mount Carmel parish, but because it is located in barangay (village) Barasoain, it is commonly referred to as Barasoain Church.

5. Note that President Noynoy Aquino's maternal grandfather and my maternal grandmother are first cousins. Due to this extended kinship relation, after the original Chichioco-Conjuangco couple moved and established their family in Tarlac (a landlocked province north of Bulacan), my grandmother became the caretaker of the Chichioco-Conjuangco ancestral home for most of her life; thus, my mother and her siblings and sometimes their children (my cousins) also lived in the large *bahay na bato* (stone) home. The home is technically on Chichioco land (and I am a descendent of the Chichiocos), but due to patriarchy and the Conjuangco family's significant wealth and political power in the Philippines, the dwelling is often seen as primarily Conjuangco property. See Tiongson (2011) for Chichioco-Conjuangco family history.

6. *Kamistisuhan* literally means "where the Mestizo/as are located, where they reside."

7. Note that Tokyo-Yokohama is first, Jakarta second, Delhi third, Metro Manila fourth, and Seoul-Incheon fifth. For the sake of comparison (in North America): New York has a population of about 8.5 million people; LA's population is approximately 13 million people; and Greater Mexico City is recorded as 21.2 million.

8. Malolos, on average, is located at 64 feet above sea level. However, villages in coastal Malolos villages are just above sea level.

9. In a relevant trans-Asian discussion, Marran critiques the use of a narrow Japanese nationalist biotrope (e.g., cherry blossoms) and notions of Japanese ethnic homogeneity used by writer Haruki Murakami after "3.11" (the earthquake, tsunami, and nuclear accident disaster that occurred in Japan in 2011). Marran suggests that writer Ishimure Michiko's more planetary and posthumanist orientation and sense of care for a "wounded earth" is ultimately the ethics and aesthetics that are needed in our toxic world. Probyn offers an excellent posthumanist, feminist, queer, and Deleuzian ethnographic account of the global seafood industry where she stresses human-fish entanglements.

PART III

Cultural Currencies

CURRENTS MOVE. They circulate, inundate, ebb, and flow. Currencies delineate and differentiate. They also unite. They bring societies into material and symbolic contact with the energy stores they possess, the values and wants to which they aspire (money, time, life, belief, resource, rule), and the resources they can expend and need to preserve. Circulating, currencies determine discourses of interchange, and as they do so, they index, activate, and define culture (Lee and LiPuma 2002). The three chapters in this section focus on the multiple meanings suggested by the idea of "currencies" when culture leads discussions of value in water policy and management strategies. Authors in this section re-endow the idea of currency with its literal meanings: circulation, flow, and acceptance. These essays conclude the volume, inviting reflection on all nine chapters as a collective hydrohumanities body of work.

By way of the hydrohumanities, scholars in this section stake claims on currencies that subvert the term's commonplace economic, capitalist, and remunerative material value. This body of study recalls a historiography of cultural critique from the nineteenth century forward: from the socioeconomist Thorstein Veblen (1899), who described capitalist economics as a predatory system of control realized through class structures built upon behaviors that establish social status over others, to the dialectical-historical materialism of twentieth-century critical theory, and into new materialisms emerging in the current century (e.g., Jay 1996; Coole 2013; Lettow 2017; Cotter, 2016; Ballestero 2019). Even today, when humanities scholars are defining posthumanist modes of inquiry, problem-solving. and exchange (e.g., Neimanis, Åsberg, and Hedrén 2015), capitalism and the cultural institutions it has borne, including practices that exhibit dominance over nature, still carry significant value in the undercurrents of critical social discourse. Considering human and nonhuman natures together as "naturalcultural" (Neimanis, Åsberg, and Hedrén 2015, 82) is becoming a commonplace imperative, proposed in the interest of renegotiating currencies of predation toward economies of integrated cultural circulation.

These authors propose a heightened focus on humanities scholarship that accelerates inclusion of culture—broadly defined—into public policy and management of water. They provide evidence for past, present, and proposed futures and offer promise for resolving current environmental problems that pose uncertainty with respect to water futures, advocating for strategies based more on cultural interrelationships with water than on dominance over nature.

At the level of a localized region—the state of California—Rina C. Faletti focuses in chapter 7 on water problems made visible by photographers. The images under discussion engage potent issues introduced on the one hand by hydrological science and, on the other, by agricultural fieldwork, revealing social inequities and cultural values related to large-scale industrial water supply development. Faletti anchors her analysis on a now-historic photograph, published in 1977 by a hydrogeologist, whose circulation ultimately influenced policy changes grounded in restructured thinking about groundwater use and water security. She connects this image to a history of agricultural documentary photography in California: "From images of transient Dust Bowl and Great Depression workers, to photographs of Mexican braceros, interned Japanese farmers, Black and Filipino migrant laborers, and United Farm Workers, photographers exposed a previously invisible social landscape and the faces of the laboring classes who bolstered massive-scale water extraction schemes." These images have unveiled previously unseen water-related crises, from geological to social, exposing them as cultural problems, and leading to mindful policy changes: "The photographic image opens critical questions onto a host of scientific and cultural concepts at once, questions that lead to action through exposure and discussion of the once-invisible problem."

At the regional and national level in the Indus Valley, James L. Wescoat Jr. and Abubakr Muhammed show in chapter 8, through a historical and systems management analysis, that, from the beginnings of human interaction with waters in the Indus Valley, water management has been most healthy when aligned with the cultural form of the irrigated garden. Yet, more than a century of technocratic water management in the region has led not only to inefficiency, but also to "unfair competition, unbridled capital accumulation, and oppressive power relations." The authors advise reincorporation of the garden idea into future policy thinking, so that "in place of the illusory goal of optimizing economic development or romanticizing traditional landscapes, the cultural approach espoused here helps articulate the 'purpose' of a river basin's existence, the 'meaning' of its people's aspirations, the moral dangers of mismanagement, and the aesthetic prospects of the irrigated garden model." This movement toward cultural models promoted by transdisciplinary humanities research and toward "jointly humanistic and scientific ideals," argue the authors, will better meet regional water needs in this region.

On the global scale, Veronica Strang in chapter 9 examines her experience as an invited expert involved in crafting the 2017 United Nations Principles for Water, a process that included explicit delineation of cultural aspects of water. Defining water's cultural uses reached beyond limited characterizations of water "culture" as ornament and recreation, for example, to consider water a key to environmental well-being. Strang argues that "it is in bridging these gaps, between different ideas about value, that the theoretical frameworks and the cultural translation provided by the social sciences and the humanities is vital." She evaluates how water-as-culture should bear upon policy-making decisions that guide global practices of water use and management into the future, observing that "deeper cultural values are not readily quantified. Social scientists and humanities scholars bring to the table robust qualitative methods that can make more complex values visible and comprehensible."

The outcomes suggested in this section necessarily involve increasing the demands and risks in the emerging realm of the hydrohumanities, as humanists reexamine the roles and responsibilities of environmental scholarship. The essays in this section, and indeed in this entire volume, ultimately propose that the hydrohumanities, in centering on water, can contribute to articulating problems and to inspiring creative solutions toward more inclusive water currencies.

REFERENCES

Ballestero, Andrea. 2019. *The Future History of Water*. Durham, NC: Duke University Press.

Coole, Diana. 2013. "Agentic Capacities and Capacious Historical Materialism: Thinking with New Materialisms in the Political Sciences," in *Millennium: Journal of International Studies*, 41 (3): 451–69.

Cotter, Jennifer. 2016. "New Materialism and the Labor Theory of Value." *the minnesota review* 87: 171–81.

Jay, Martin. 1996. *The Dialectical Imagination: A History of the Frankfurt School and the Institute of Social Research, 1923–1950*. Berkeley: University of California Press.

Lee, Benjamin, and Edward LiPuma. 2002. "Cultures of Circulation: The Imaginations of Modernity." *Public Culture*, 14 (1): 191–213.

Lettow, Susanne. 2017. "Turning the Turn: New Materialism, Historical Materialism and Critical Theory." *Thesis Eleven 2017* 140 (1): 106–21.

Neimanis, Astrida, Cecilia Åsberg, and Johan Hedren. 2015. "Four Problems, Four Directions for Environmental Humanities: Toward Critical Posthumanities for the Anthropocene." *Ethics and the Environment* 20 (1): 67–97.

Veblen, Thorstein. 1899. *The Theory of the Leisure Class: An Economic Study of the Evolution of Institutions*. New York: The Macmillan Company.

7

The Invisible Sinking Surface

Hydrogeology, Fieldwork, and Photography in California

Rina C. Faletti

HYDROGEOLOGY AND A PICTURE
OF A SINKING STATE

In 1977, United States Geological Survey (USGS) hydrogeologist Joseph Poland staged a symbolic photograph (figure 7.1). Standing by a power pole in a typical California Central Valley agricultural landscape surrounded by vineyards, Poland posed without fanfare, in shirt sleeves, khaki pants, and a hat, one arm up against the pole, the other down by his side. Tacked on the pole at varying heights, from the ground at his feet to high above his head, four large placards printed in bold black letters delivered a cryptic environmental message. The highest placard, thirty feet off the ground, displayed the year 1925; halfway down the pole, another announced the year 1955; and, leaning against the pole on the ground near Poland's feet, a third identified the current year of the photograph, 1977. A fourth sign, attached at about head height next to Poland's uplifted arm, listed five facts in block text:

> *SAN JOAQUIN VALLEY*
> *CALIFORNIA*
> *BM S661*
> *SUBSIDENCE 9M*
> *1925–1977*

With this, Poland delivered his strident visual message: in the San Joaquin Valley of California where he stood, at the topographical benchmark numbered S661 near the town of Mendota, the earth's surface had subsided nine meters, nearly thirty feet, between 1925 and 1977. Translated, Poland and his USGS team that devised the photograph conveyed its urgent report: that in the course of fifty years the land's surface where Poland planted his feet had sunk an average of a half a foot a year, and was still sinking (map 7.1).[1] A newspaper article from the period put it

FIGURE 7.1. Hydrogeologist Joseph Poland, USGS, standing at the point of most extreme land subsidence near Mendota in the San Joaquin Valley of California, 1977, by Dick Ireland. James L. Borchers, *Land Subsidence 1998* (USGS 2019), 1, 65. Photograph is in the public domain. File provided courtesy USGS Water Science Center, access provided by Sally House, Science Communications: www.science-base.gov/catalog/item/58335611e4b046f05f21f69f.

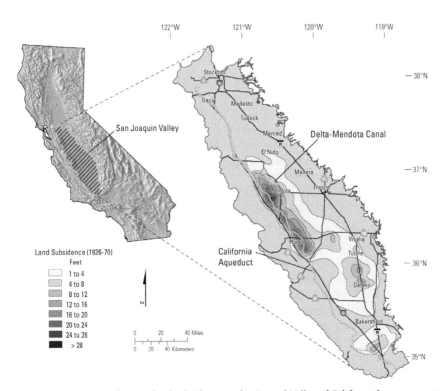

MAP 7.1. USGS Map showing land subsidence in the Central Valley of California between 1926 and 1970. Note the point near Mendota, Poland's Benchmark Number S661, the point of most extreme subsidence. Online in meters at http://water.usgs.gov/ogw/pubs/fs00165/ (Land Subsidence in the United States, USGS Fact Sheet-165–00). Public domain map provided courtesy USGS California Water Science Center, prepared by Michelle Sneed, Hydrologist.

this way: "Joe Poland stands beside a utility pole that dramatically demonstrates how far land . . . has sunk in 52 years. Without subsidence, the grapes to the right would be growing at the 1925 land level" (Barnes n.d.). Poland's unassuming pose invited viewers to stand in his place, crane their necks upward more than thirty feet along the power pole, and imagine their own two feet standing upon a now-vanished land surface that had hovered high above their heads in the empty air. At the same time, the image unveiled the invisibly subsiding water stores beneath Poland's feet, exposing the patent fact that his soles were not planted upon terra firma, but upon a steadily subsiding land surface. Poland's visual message sounded an alarm, that the time had long passed to remedy California's crisis of disappearing groundwater.

With the 1977 photograph, Poland and his USGS groundwater team had devised a simple visual means of conveying a complex hydrogeological concept: *land subsidence* is the sinking—or subsiding—of an extent of the land surface, resulting from groundwater pumping through wells. As water is pumped to the surface, the subterranean spaces compact and compress, and the land surface sinks with them.

With excessive pumping from large-scale wells, compaction of porous clay-rich soil layers (aquitards) reduces the spatial volume of aquifers (underground water stores are held in aquifers and aquitards) as internal water pressure reduces. Without the internal water pressure, the soil and rock that make up the aquifer structure cannot bear up under the structure's own weight: the aquitard spaces collapse, and the land surface capping them subsides with it. In other words, the "surface expression" of subsurface collapse and compaction is land subsidence (Prokopovich 1973, 191). USGS California Water Science Center hydrologist and land subsidence specialist Michelle Sneed explains the hydrogeology this way: "All sediments are porous to some degree, but aquitards 'retard flow' because water doesn't flow very well through clay-rich material. Put water on top of playdough and see how long it takes to flow through (it'll evaporate long before it flows through)."[2]

Try another analogy: fill up a box with full water balloons. Balance a substantial flat object—let's say a geology textbook—on top of the balloons so that the book is supported by the water pressure held by the balloons. The balloon surfaces and water under pressure within them represent the subterranean aquitards and aquifers holding the water in. The geology textbook is the land surface. Now, release a little water from each water balloon, one by one, so that the pressure the water exerts on its surface diminishes. Without the water's pressure keeping it filled, the balloon's material surface compacts, and the spaces that once held water close in on themselves. The balloons collapse and flatten, and the geology textbook begins to sink, at a rate of subsidence equivalent to the water pressure's release. With the withdrawal of water from the aquifer, the aquitards compact, and the land surface—the textbook—ends up at the bottom of the box, or, in the visual language of this chapter, beneath Joseph Poland's feet.

At the time of Poland's photograph, the extent of sinking land surface in the Central Valley amounted to more than forty-three hundred square miles affected, with one to thirty feet of vertical land subsidence taking place between 1926 and 1970 (Poland et al. 1975, 11) (map 7.1). And, it is still subsiding.[3] Today, the USGS webpage displays Poland's photograph with an annotated caption that summarizes the land subsidence science behind the balloon and playdough analogies:

> The compaction of unconsolidated aquifer systems that can accompany excessive ground-water pumping is by far the single largest cause of subsidence. The overdraft of such aquifer systems has resulted in permanent subsidence and related ground failures. In aquifer systems that include semiconsolidated silt and clay layers (aquitards) of sufficient aggregate thickness, long-term groundwater-level declines can result in a vast one-time release of "water of compaction" from compacting aquitards, which manifests itself as land subsidence. Accompanying this release of water is a largely nonrecoverable reduction in the pore volume of the compacted aquitards, and thus a reduction in the total storage capacity of the aquifer system. (USGS 2018)

Poland's role as photographic courier had been percolating for years. His first geology field research in the 1930s as a graduate student at Stanford University identified the new scientific reality of land subsidence as measurable and caused by subsurface water drawdown across entire valley regions. Continuous research through subsequent decades proved that subterranean soil compaction and a sinking land surface followed from a persistent practice of industrial pumping of groundwater for irrigation; his projects had also concluded that withdrawal of petroleum caused not only land subsidence but also sea water intrusion into fresh water aquifers near the coast. In 1954, Poland joined the USGS in Sacramento to lead the first major study of land subsidence in the United States. There, he changed the scope of his profession in two ways: first, by combining the formerly separate fields of land and water study to create the discipline of hydrogeology, and second, by initiating on-the-ground research in the field as part of scientific study. Rethinking the traditional practices of modeling derived by mathematics computation and of in-lab study, Poland "perceived that areas of land subsidence provided natural field laboratories" (Riley 1998, 19). He reasoned that, working outdoors in the field, hydrogeologists could gain "more comprehensive understanding" of "difficult-to-measure but vitally important in-situ properties" of interactions among land, water, and grand-scale industrial pumping than they would derive from in-lab and mathematical study only (Riley 1998, 19–20). In the 1970s, Poland began taking colleagues on field trips to the point of most extreme subsidence near Mendota, California, where he and several of his USGS hydrogeology colleagues conceived and staged the 1977 photograph (Borchers 1998, 64; USGS 2018; 2019).

The photograph achieved two aims, one informational and another experiential. First, the image made land subsidence clear in physical terms, and second, it delivered a shock experience of discovery, realization, and urgency (Shiff 1992). With this, the photograph succeeded—and still succeeds—in making viewers feel complicit in the act of inflicting irreversible damage upon the earth, at least in those viewers disposed, like Poland, toward empathy regarding environmental consequences of natural resource overextraction. Within the groundwater world, the man by the pole is recognized as Joseph Poland and makes an indexical reference to the scientific work of the USGS. Outside that world, for viewers who do not realize the message is related to the identity of the messenger, the anonymous figural image (man, ground, pole, placards) signifies to any audience—from water experts to policy makers and the general public—the physical effect of groundwater pumping, the sheer scale of a vanished earth's surface, and the importance of invisible underground bodies of water permanently lost. The body of scientific, hydrogeological knowledge informing Poland's deceptively simple photograph exposes the invisible consequences of water resource extraction and its environmental effects, and as it does so, it sharpens an impulse in viewers to question that situation, to interrogate the powerful forces in industry, technology, and capital that leveraged resources to create a modern California. An entire future, now past, has funneled water, land, and people as if these were currencies due on a promissory note banking on what water can grow. The photographs ask society to address a nagging cultural uncertainty about the state of the environment, as a way to articulate the question now emerging, of how to live *with* water into a next, more stable, future.

The invisibility this chapter addresses takes two forms: one of land and the water within it, another of the laborers who work with that land and that water. At the outset and on the surface, the chapter analyzes land subsidence: the measurable but invisible sinking of the earth's surface with subterranean soil compaction due to agricultural groundwater overdraft over time. This is physical subsidence. The chapter also interrogates a hidden subsidence of a social kind: the socioeconomic and racial leveling of agricultural farmworkers who operate beneath a figurative cultural surface. Both types of subsidence—physical and social—are historically connected to the extremes exacted by industrial agriculture's water uses. This chapter examines ways in which mindful photography created a field of visual problem-solving regarding water, land, people, and ideas that labor on and with it. An interesting distinction appears in the images I analyze later in this chapter: in the photographs of the scientist, the photographer, and the organizer, their *professional* bodies are presented as discrete from time and removed from the land they study and capture, and upon which they gather. Contrast this with the photographs of farmworkers, whose *bodily labor* and the land they work are intertwined and inseparable. Juxtaposing and paralleling photographic representations of worker, land, and water, with an eye toward how they are enmeshed in their difference, can reveal changing ideas about how water *works* and how it *is worked*, on and under the surface.

VISUALIZING SUBSIDENCE: MODEL AND PHOTOGRAPH

Joseph Poland's photographic representation of land subsidence succeeded in staging the conundrum of placing the invisible on view. Poland's image did this in the way it presented the forms in the picture, in effect both reinforcing and defying standards for typical landscape photography, typical portrait photography, and typical agricultural photography of the era. He also built upon a practice by scientists of his time, to present scientific data in visual form, usually by way of physical, built models.

The landscape in which Poland stands is recognizable as a California agricultural landscape, with vineyards extending to the horizon on a flat extent of land. The vineyards are bounded, divided, and traversed by a road and lines of power poles. These appear vertical in the photograph, in lines that reach to or parallel the horizon, though the horizon's vanishing point is blocked by a farm complex of buildings in the background above Poland's head. The hydrogeologist stands in the landscape ostensibly as a measure of comparative scale, in ways that depart from the expected role of the human subject in traditional landscape painting and photography and in the expectation set up in documentary agricultural photography (Borchers 1998, 64–65).

The extreme vertical orientation the pole creates, joined with the proximity of the vertical elements to the "front" of the picture plane, requires the "foreground" to extend upward to the top of the picture plane. This permits a direct focus on the "1925" year sign, where the electrical power lines converge. The power lines anchor *both* the flatness that the pole's verticality imposes in the foreground *and* the illusion of receding poles moving toward the blocked horizon. These effects are amplified by the aerial perspective: the camera is positioned above the scene, looking down upon it. All of this remains undetected by the illusion of simplicity in the photograph: a man stands next to a pole in a vineyard. In this way, the picture succeeds in linking the vertical to the horizontal, allowing the vertical plane to represent an enigmatic "depth" that inverts the traditional idea of landscape depth, usually defined by the recession that perspective creates, toward a vanishing point on the horizon. I say "inverts" because now, instead of the landscape picture indicating depth as the quality of optical recession in a picture of a landscape, it reconfigures depth vertically, with the pole as a sign for volumetric depth: of water under the earth. In other words, the pole can also be read as a well shaft, both exposed by past subsidence and descending underground into the invisible terrain yet to be unearthed by the sinking ground surface upon which Poland stands.

Poland had been working for years on ways to represent the invisible depths of land subsidence in visual form. Grounded in his early work in the Santa Clara Valley in the 1920s and '30s, where he and his research team confirmed and named the phenomenon of land subsidence, Poland became a pioneer in creating visual representations, models, and photographs that would allow scientists, and the general and governing public, to understand it. As early as 1932, Poland was sharing with colleagues his finding that the field of groundwater geology was "of vital importance in the future development of California. When one realizes that at present, the annual drop in subsurface water level averages from four to six feet in most of the California valleys, the necessity for artificial replenishment in the near future is readily evident" (Poland 1932).[4] This was a "visionary" concept in 1932, according to Sneed, but is a growing necessity today.[5] Poland included a historical caveat, that little work could be done in his time on the issue of groundwater, either by private individuals or by public agencies, due to "the state of business" wrought by the Great Depression, adding that "primarily for that reason, we have barely been keeping the wolf from the door"[6] (Poland 1932).

Hydrogeology reports on groundwater and subsidence came to rely on two-dimensional graphs, drawings, tables, charts, and maps, all standards in scientific publication (e.g., Curtis, Reid, and Ballard 2012; Latour 2014; Pang 1997; Perini 2006; Roberts 2017; Rudwick 1976). For public and nonspecialist display, Poland devised three-dimensional models for newspaper articles and presentations when he reported on land subsidence. This followed a period standard: a scientist or engineer stood in a photograph next to a physical three-dimensional model, with a caption describing the model and the concept the model represented. The model-author expert substantiated the science. Poland's USGS graphic models and his clear claims, backed by visible scientific evidence, contributed to newspaper reporting on land subsidence findings. In Southern California, articles and photographs on the topic ran in several newspapers (Gesner 1941; "Heavy Water Pumping..." 1946; "All Water Wells Pegged" 1947; "Water Conservation Urged..." c. 1940s–1950s). In the 1940s, newspapers related environmental consequences of groundwater depletion and salt water intrusion resulting from overpumping of water or petroleum near coastlines. In conjunction with reports on a USGS survey of land subsidence and salt water infiltration along California's coast, photographs appeared of the subsidence model Poland and the USGS had devised, representing land surface levels sinking to below-sea-level elevations. The model took the form of a tabletop punctured with vertical pegs of different lengths, each representing a well. The tabletop to which the pegs were attached indicated groundwater levels, and a string tied horizontally above the tabletop model represented sea level. The aim of the model was to demonstrate that the ground level had sunk below sea level. One of the first photographs of the peg model appeared in 1941, with Poland and geologist A.M. Piper standing next to it. The Los Angeles *Daily News* article, on sea water infiltration into freshwater aquifers due to groundwater overdraft, included the following description of the peg model:

> The USGS has developed an interesting method of charting drilling logs of the thousands of wells involved. A three-foot shaft—resembling the shaft of an arrow—is marked in various colors along its length. The colors represent the differing formations encountered in drilling. Thousands of these shafts, each representing a well, are stabbed into a large map at the site of the well, giving an upward projection of the various formations underneath. This enables the engineers to make a three-dimensional study of the underlying formations. (Gesner 1941)

The power pole in Poland's picture, then, might represent an exposed well casing, as if from the peg model writ large directly upon terra firma. Sneed reports that this literal perception is not uncommon: she has often been asked at scientific meetings whether the utility pole in Poland's photo might actually be a well casing made visible by subsidence.[7] In Poland's picture, the topmost, "1925" placard signified the now-invisible past land surface *and* the volume of its prior subsurface depth, from aquifers and aquitards long since emptied, compacted, and vanished. Where the peg model would also show the subterranean well peg below the surface, Poland's viewers were left to the power of suggestion, "seeing" by association the sub-depths beneath the subject's feet, now (before and since 1977, and, without remediation, forever) in the process of disappearing. The picture enacts an ongoing future of subsidence.

The photograph imparts information, but its legacy is that it also ignites a shock experience (Shiff 1992). The *experience* of the photograph is to imagine standing on the ground's past surface, "up there" in the empty space above the pictured man's head: above our heads. This brings a sense of the *inversion* of the electrical pole: it both rises from and punctures down into the ground where it stands. The pole is as much a utility pole rising up from the ground in 1977 as it is a likeness of a well shaft bored into an aquifer in 1925. When this well shaft analogy stretches the reach of the image, allowing the pole to signify the invisible past store of water whose disappearance has taken place in an unseen subterrain, then the pole-as-shaft marks the site of a puncture, a wound both hidden and exposed. The experience brings this insight, that human industry physically wounds the earth, and confirms an ethical weight to the problem, as viewers see that, being part of the industrial culture that extracts water without limits, we are complicit in the breach and the wound. The ultimate power of this image is its requirement that we—its viewers—engage the grave realization—the shock—that we are accessory to and yet beneficiary of this continual, lasting and irreparable harm. Moving further, attuned viewers see that this is not a "past" situation, but a recurring, and current, problem: in 2021, we are nearly fifty years past 1977, and approaching the century mark from the 1925 ground level. Subsidence continues. The experience of Poland's immersion in a field of absent water brings new knowledge, that the effects of subsidence are wounds—punctures, fissures, tears, cracks, depressions, sinkholes—in the earth's surface as an entire region physically sinks and keeps sinking. When the photograph "makes sense"—when the viewing becomes a sense experience of a physical harm that results in the earth's very movement—it brings associative moral pain, guilt by cultural association, shame of the unwitting accomplice (Benjamin [1931] 1977; Benjamin [1936] 1968). Apprehended fully, the photograph activates a sense of collective responsibility for the pain of a wounded planet: these are wounds we helped to cause, against earth, against water, against ourselves.

PHOTOGRAPHIC FIELDWORK AND PHOTOGRAPHS OF FIELD WORKERS

Between the 1920s and the 1980s, out of the growing interest in the social inequities that accompanied water systems development in California grew the field of social documentary photography. From images of transient Dust Bowl and Great Depression workers, to photographs of Mexican braceros, interned Japanese farmers, Black and Filipino migrant laborers, and United Farm Workers, photographers exposed a previously invisible social landscape and the faces of the laboring classes who bolstered massive-scale water extraction schemes (e.g., Finnegan 2003; Gordon and Okihiro 2006; Lange and Taylor 1939; D. Mitchell 2012; Street 2004; 2008). The rise of California's network of water conveyance systems during this time, both across and beneath the ground's surface, participated in an industrial-scale agricultural empire meant to feed and water California's insatiable twentieth-century regional metropolis far into the future. Poland's career in

California groundwater study, which began in the 1930s during the Great Depression, was concurrent with the historical timeline of California's large-scale industrial agricultural business. By association, that career also grew up alongside the social history of labor groups that the agriculture industry relied on to make that business productive and profitable, and it bore witness to the social justice movements that worked toward equity for those workers. Paralleling this history came the social documentary photography profession that recorded it all in pictures.

In the foreword to Richard Steven Street's book *Photographing Farmworkers in California* (2004), California historian and former state librarian Kevin Starr (2004) observed about California's agricultural lands that "the pastoral contentment of the early photographs—their emphasis on field, sky, crop, and willing workers—now yields to photographs documenting the remorselessness of piecework in the fields, the inadequacy of housing, the early efforts at organization, and the browning of the workforce" (Starr 2004, xiii). This identifiable conflict between art and documentation in agricultural photography arose in the 1930s, when photographers began to shift from pastoral views of field labor, inherited from more traditional landscape traditions, to social commentary focusing on race and class injustice (Goodwin 1998; Starr 2004). This shift, with photography emerging as witness and social critique—even as protest—suggested a struggle within "the interaction of documentation and aesthetics, especially when art is offering, or so it seems, a shortcut to a more fundamental truth" (Starr 2004, xiii). Indeed, the scientific and social documentary photographs I discuss in this chapter attend to Starr's "more fundamental truth" about historical relationships among water, land, and people in California (2004, xiii). Here, the material value of water rose as it was abstracted into economic and social capital, lost its physical properties, and became a commodity prone to be considered waste; the social status of the laboring bodies that sustained this system subsided in status as work and worker became commodified (Cotter 2016; D. Mitchell 1996). This seeming conundrum harks back to the pioneering transparency of social economist Thorstein Veblen (1899), who made plain that *status*—claims of social superiority based on the appearance of material wealth—creates high social stability in a privileged class and requires the judgment that people outside the self-proclaimed class of privilege are lower in value. The tendency to obscure, then, in the context of Central Valley water history, not only hides geological problems; it also curtains consequences of the-problem-that-must-not-be-named to the environment and affected social groups. Over time, nested cultural practices worked to keep water problems, and their social and ecological effects, invisible. The "more fundamental truth" that underlies the photographs, then—which Starr leaves unnamed—is a truth of unjust and uncorrected practices that drive toward social imbalance, when water is a capital companion to land and agricultural development (Starr 2004, xiii).

Writing about Depression-era photographs of Black American laborers, James Goodwin insists that "mere depiction or description of appearances does nothing

to lift the veil" from the deep cultural norms governing social thought and practices about status in relation to skin color and social class (Goodwin 1998, 284).
Analyzing Richard Wright's text for his book of social documentary photographs,
12 Million Black Voices: A Folk History of the Negro in the United States, Goodwin
traces ideas of race in American society from W.E.B. Du Bois to Wright, finding that "one psychological effect of the color line in America . . . [was] to have
drawn a veil that screens the black world from the white one. For Black Americans
this veil, depending upon the social context, proves to be variously impenetrable,
reflective, or, at moments, translucent. For white Americans, the veil is presumed
to confirm stereotypes or phantasies about blacks or to attest their invisibility"
(Goodwin 1998, 284; see also Frank 1998; Puskar 2016; Shiffman 2007; Street 2004;
Starr 2004; Wright and Rosskam 1941). In looking at agricultural social documentary photographs of nonwhite field workers in California, the practice of obscur-
ing race and class "through a calculated opaque mask" (Goodwin 1998, 284) is a
patent race-class invisibility grounded in long-held cultural norms (D. Mitchell
1996). Social documentary photographers in California from the 1930s onward
unveiled problems of an imposed marginalization in their pictures of bodies laboring within the land. What were the margins, and who was pushed into margins
by whom, for entire social groups of field workers? In a context of geographical,
social, economic, and political segregation, a lowered quality of life—low status—
played out in all its racialized variations: in control by forced labor and erasure of
native Indigenous peoples; in generations of Chinese laborers beginning with the
Gold Rush and the Chinese Exclusion Act in the nineteenth century; in waves of
Dust Bowl and Depression-era migrants in the 1920s and '30s; in the internment
of Japanese at war relocation camps during World War II; in the bracero workers
cargoed from Mexico to replenish a shrinking labor force; and in the Black, Filipino, and Mexican migrant field workers who subsequently created the context for
the United Farm Worker movements into the 1970s.

Poland's own fieldwork was taking place along the duration of these movements. He had joined the Sacramento office of the USGS in the mid-1950s to
lead newly established research projects on groundwater and subsidence science.
During his entire tenure there, farmworkers were organizing throughout the San
Joaquin Valley and California, in many of the same locations as his field laboratories. Most prominently at that time, nonunion Filipino and Mexican field workers
comprised the backbone of California's early United Farm Worker movement, led
famously by César Chávez and Dolores Huerta (Gunckel 2015; Street 2008). The
organized strikes, marches, and fasts, and the attendant violence, were magnetic
subjects for social documentary photography. To this day, prominent social documentary photographers like Matt Black, for example, raised in the Central Valley,
focus the lens of exposé on the repeat-shock of continued race-class depression
and social injustice in Central Valley agricultural and rural communities. Jason
Puskar has argued that photography has historically occupied two territories:

"deeply complicit with Western racism" but also able to "disrupt entrenched power relationships" (Puskar 2016, 169). Agricultural documentary photography continues to play its role in exposing issues of race, gender, and class while "shaping public perceptions of life and labor in the fields" (Street 2004, 300; 2008).

SOCIAL STATUS, RACIAL ERASURE

In 1935, Dorothea Lange, the New Deal photographer, in her most famous portrait, positioned herself on the roof of a car, balancing her camera upon her bent knee and peering through the viewfinder to a scene outside the picture frame (Taylor c. 1935). For documentary photographers, shooting from the top of a car was a common way to gain the elevated vantage point that could capture the people working in the flat extent of the Central Valley landscape. In one way, then, this photograph of Lange documents a photographer's method, but Lange's portraitist stands at ground level shooting upward. This point of view accentuates Lange's separateness from the territory of the farmworker. Her body, like Poland's, unites physically, through touch and positioning, with the tools of her documentary trade (car and camera). Lange's photographer accomplishes this fusion formally, with the triangulated compositional form of her pose upon the car. And, looking up at her from the ground, her portraitist captures her in concert with car roof, high-power lines, and clouds, presenting Lange as existentially *elevated above* that which she photographs: she represents elevated status. As in Poland's picture, vanishing points and horizon are blocked, yet implied. The picture reads as a figure in a landscape. Lange's pose presents an independent, specialized, middle-class, white worker in a field whose labor, gender, and class status structure her at a distance from her subject: field laborers *within* the landscape, absent but implied, outside the frame, working *with* the land. Her role in this photograph (and in others like it from the period) is to observe and comment *upon* and *above* the land, not to participate *within* it. Like the scientist, the photographer is apart from the landscape, not a part of it.

Lange's portraitist, her partner in work and marriage Paul Taylor, was well known as a University of California, Berkeley immigrant-labor economist. The pair embarked together on a study in the Central Valley to explore and expose social, economic, and racial inequities imposed on migrant and immigrant working poor in agricultural California. Their 1939 book, *An American Exodus: A Record of Human Erosion*, is a photo-text that exhibits "apprehension over the adequacy of photography to convey the full complexity of the social changes it examines" (Goodwin 1998, 287). Certain of Lange's Farm Security Administration (FSA) images had cemented her identification in the public eye with the human crisis of Dust Bowl drought on the American Great Plains, but for this joint project, she and Taylor worked more in California's agricultural terrains. There, Lange immersed herself in agricultural social structures and practices to carry out both

her FSA work and the collaborative projects with Taylor. Combined, the two projects exposed collapse within the long-held rural farming ideals that were still
active in the American cultural imagination. As Linda Gordon explains:

> The FSA's photography project was supposed to promote not only Department of
> Agriculture programs but also a New Deal vision for rural America, a difficult as-
> signment because of the incoherence of that vision. The project reaffirmed family-
> farm ideology through its frequently romantic, picturesque approach to a "simple"
> and community-spirited rural life and its condemnation of plantation and industrial
> agriculture. . . . [Yet], the extraordinary popularity of some of [Lange's] photographs
> has decontextualized and universalized them, categorized them as art, and thereby
> diverted attention from their almost social-scientific significance. (Gordon 2006,
> 700–701)

The photographs of Dorothea Lange and Joseph Poland at work on the landscape
present them as confident, modern, master professionals fused with the tools of
their technological wizardry. Each is an expert observer and recorder working *upon
and over* the land, not as laborer *within* it. Contrast these images with a famous
photograph of Dolores Huerta, the principal organizer, with César Chávez, of the
United Farm Workers movement. The photograph was taken by documentary
photographer Harvey Richards, who titled it *Dolores Huerta HUELGA Sign, Without a Doubt the Iconic Photo of the Delano Grape Strike, 1965* (Richards 1965).
Huerta, like Lange, stands atop a car facing toward the farmworkers with a sign
marked by the United Farm Workers logo, declaring "HUELGA: STRIKE." Like
Lange, Huerta stands apart from and above the field workers she faces. She is fused
to the material of her work as well: the emblematic sign she holds and the car on
which she stands.

All three photographs of these professionals—scientist, photographer,
organizer—are backed by power utility wires, which create a formal series of
lines that function visually as a background. The background serves to anchor
the subject, and it also helps move and settle the eye around the picture, but in
this case, the high power lines also convey a subtle message of the *power behind*
the figure. Differences between Huerta's portrait and that of Lange or Poland are
many. Unlike the scientist and documentary photographer, the union organizer's
face and posture present an active, emotional, engaging participant: she faces the
farmworkers and the field with a direct and commanding form of communication
meant to actively support organization, leadership, and action. Differently from
Lange's internally gathered form and Poland's poised frontal stance, the energy of
Huerta's body opens outward in a bold gesture of inclusiveness toward those she
invites into the act of protest and change through mass social movement.

Yet, in these portrait-like photographs of observer-professionals in the field
who studied, photographed, or organized, the ever-present but elided subject is the
farmworker. The observer-professionals are characterized as *apart from* or *above*
the landscapes *on* which they worked; at the same time, the laboring field workers'

bodies, to which this discussion turns now, are represented as *a part of* and *within* the land they work. Mitchell attributes this disequilibrium to industrial ideologies whose tenets of operation kept farm labor pauperized, racialized, and imbued with inferiority—that is, poor, of color, and powerless (D. Mitchell 1996, 1–12).

Now, consider Dorothea Lange's 1938 photograph *Mexican Grandmother of Migrant Family Picking Tomatoes in Commercial Field.* Here, a laboring figure appears inseparable from the landscape in which she works (figure 7.2). Held down by a horizon line that seems to press down upon her back and her tightly curved body, she is multiply and inescapably anchored to the earth. Her feet, hidden behind her picking bucket, sink into the muddy furrows, steadying her as her hands disappear within the gnarled tomato vines she reaches to pick. Her black bucket and black head scarf anchor her to the field, the curved masses also working formally to keep the bent body compressed into its tight stoop. She is an expression of the stoop labor that became the symbol for and target of farmworker protests. The vanishing point, that typical hallmark of traditional Western land-scape art that pulls the eye to a far horizon, is in this case pushed to the lower left corner of the picture frame, as if by the pressing of the field worker's foot against the bucket's weight. Here is a laboring body whose identifying features are mute: faceless, sexless, nameless, she strives alone with the earth and with the product of her work in "the commercial field." Lange presents a persuasive portrait of the totality of commodified labor, a documentary shot taken at the same time and in the same location where Joseph Poland's Stanford team proved land subsidence to be the result of groundwater pumping. Their research was under way in this very area, the Santa Clara Valley, when Lange captured the image in November 1938 (Tolman and Poland 1940).

Similarly to the land-bound tomato picker, documentary photographer Paul Fusco's field worker subjects toil under the weight of full picking boxes, heavy enough to make their boots sink into the tilled soil between vineyard rows (figure 7.3). *La Causa: The California Grape Strike, Farmworkers Carrying Grapes* (1968) captures the blur of action as grapes fly from the crate into the fruit collection bin. The two workers' facial expressions draw viewers into the physical labor that speeds at the pace of a by-the-box pay scale. Fusco has captured the urgent efficiency with which these pickers must do their jobs. They trudge within the vineyard rows, their hands, faces, and bodies soaked with sweat, caked with dirt, sticky with juice, and filmed with dust, pesticide, and bugs. The camera frames them as part and parcel of the soil and the plant rows themselves: they are the same height as the loaded, mature vines; their heads are even with a horizon indiscernible from the planted terrain itself. Gathered into the thick vine foliage at the front of the picture plane, vanishing points truncate where they meet the laboring bodies—handles angle toward the collection bin, slanted shadows of legs dig into the foreground, diagonal lines of the furrows hide behind their muscle weight.

FIGURE 7.2. Dorothea Lange, *Mexican Grandmother of Migrant Family Picking Tomatoes in Commercial Field. Santa Clara County, California* (November 1938). Photograph is in the public domain. Library of Congress, Prints & Photographs Division, FSA/OWI Collection, LC-USZ62–125640 or LC-USF34- 018409-E [P&P] LOT 346]. Online at http://hdl.loc.gov/loc .pnp/cph.3c25640.

These formal elements require the eye endlessly to cycle at the front of the picture, the forefront of action, never drawing through to the horizon. This is not a landscape, but a visual statement of the short view, on the ground, an endlessly repeated moment of immersed labor, a compression of time in specifically human

FIGURE 7.3. Paul Fusco, 1968, *La Causa: The California Grape Strike, Farmworkers Carrying Grapes*. Photograph courtesy Magnum Photos.

terms. It depicts not the agency of a professional observer of work, but the agency of a system that drives workers.

ENVISIONING INVISIBLE SUBSTRUCTURES: LANDSCAPE AND LABOR, WATER AND WORK

The reinterpretation of landscape by social documentary photographers in the 1930s focused on working bodies' relationships *within* the landscape rather than depicting landscape itself as an aesthetic object. Yet, even these photographers were vested in modern photographic practices that compelled them to make photographs noteworthy aesthetically. The federal projects of the Depression era were contemporary with the developments of socially-minded critical landscape discourses following in the footsteps of post-World War II social and Marxist economic theories, such as critical theory arising from the Frankfurt School (Jay 1996). Led by literary scholars, cultural geographers, and labor scholars, these approaches subverted traditional post-Enlightenment and Romantic concepts of landscape-as-nature and of the wilderness sublime, instead analyzing landscape images with social, economic, and political interpretations that underscored critical symbolic content. This dimension made landscape representations legible as social constructions, an idea based on a "powerful visual ideology" (D. Mitchell 1996, 2) embedded with symbolic and/or iconographic meanings that encoded the values of the societies and time frames in which they were produced

(Daniels and Cosgrove 1988). And, this approach addressed the question: who produced the construction, the idea, the ideology of landscape from the image of "a structured portion of the earth"? (D. Mitchell 1996, 2). This question bypasses the interests of individual artists, and points to patrons, collectors, and sponsors of works of art, to include government-sponsored New Deal programs, those individuals or entities with the power to commission, pay for, publish, and exhibit them. Social approaches probed and depicted political, economic, social, and cultural conditions, opening space for artists, and even scientific field-workers such as Poland, to create new ways to see landscape.

To analyze a work of landscape art—from Renaissance painting to contemporary photography—under these terms reveals that landscape traditions themselves were "integral to an ongoing 'hidden' discourse, underwriting the legitimacy of those who exercise power in society" (D. Mitchell 1996, 2; see also Bentmann and Müller 1992; Daniels and Cosgrove 1988; W.J.T. Mitchell 1994; Olwig 1993). As Don Mitchell points out, citing Cosgrove, Raymond Williams, and others: work makes landscapes; labor power is embodied; and "a landscape is a 'work'—a work of art, *and* worked land. . . . It is a produced space" (D. Mitchell 1996, 7, 10, 6). In fact, he insists:

> One of the purposes of landscape is to make a scene appear unworked, to make it appear fully natural. So landscape is both a work and an erasure of work. It is therefore a social relation of labor, even as it is something that is labored over. To ignore the work that makes landscape . . . is thus to ignore a lot of what landscape *is*. (D. Mitchell 1996, 6)

Going further, W.J. T. Mitchell advocates not for thinking about "what landscape 'is' or 'means' but what it *does* as a cultural practice," of establishing and maintaining cultural power relations through images: "What we have done and are doing to our environment, what the environment in turn does to us, how we naturalize what we do to each other, and how these 'doings' are enacted in the media of representation of what we call 'landscape' " (W.J. Mitchell 2002, 1–2).

Poland's scientific photograph fused the hydrogeologic model with the landscape it studied, disclosing the consequences of unchecked industrial groundwater pumping for agribusiness. Social documentary photographers fused the body of the disempowered worker with the body of the worked landscape. Both exposed issues of agency and equity, and both challenged viewers to take an activist position. To compare formal and iconographical aspects of these two discrete representations—scientific and social—renders the photographs open works, candidates for thick analysis (Eco 1989; Neimanis, Åsberg, and Hedren 2015). This deepens photographic analysis as a means of defining agencies of "work" and "labor" through juxtapositions of dissimilarities between images and ontologies of the two (D. Mitchell 1996). Specifically in relationship to photography attuned to water issues and California agriculture, these critical reflections reveal a "connection between the material production of landscape and the production of landscape representations, between work and the 'exercise of imagination' that makes work and products knowable" (D. Mitchell 1996, 1–2). This questioning opens fluently within the environmental humanities approach this volume advocates for the examination of water's cultural currencies.

From the 1930s through the 1970s, Poland's fieldwork throughout the landscapes of California's key areas of groundwater overdraft and land subsidence would have immersed him in the period's culture of heightened water systems development and agricultural expansion, realities accompanied by intensified farmworker struggles and union organization. After 1956, Poland "was able to devote full time to the subsidence studies," when plans for state and aqueduct projects on the west side of the San Joaquin Valley raised subsidence concerns, "both hydrocompaction and aquifer-system compaction, because of excessive ground-water withdrawal" (Poland c. 1980, 2–3). The design and placement of the California Aqueduct of the State Water Project, built in the 1950s and 1960s, had benefited from an interagency study, begun in 1954 and headed by Poland, whose research identified "optimal siting" outside the danger zone for subsidence, and the planned aqueduct was rerouted (Borchers and Carpenter 2014, 26; Prokopovich 1973). The 1954 study was motivated in part by the aqueduct's predecessor, the 1951 Delta-Mendota Canal of the Central Valley Project: major structural damage to the canal from land subsidence had been discovered then, and is ongoing now. The case of the redesign of the California Aqueduct based on hydrogeological science represents a major success in the application of hydrogeological findings to engineering developments. But this is in some ways an anomaly. Legal decisions regarding groundwater geology have not always been made on clear understandings of the science. Land subsidence pioneer C.F. Tolman's comments on the politics of law and knowledge about water are revealing. In a 1940 letter to Poland, Tolman commented on a judge's statement in a water-related court case finding: "This is an example of the difficulty of training a judge in the fundamental principles of groundwater so that he does not invent some imaginary condition."[8]

After the California Aqueduct opened, the increased supply of surface water led to a decline and eventual halt to Central Valley groundwater pumping for several years, until demand once again outpaced supply and pumping resumed (Poland 1984; Poland and Lofgren 1984; Sneed, personal communication). Through the decades and into his retirement in the 1980s, Poland worried that aqueduct and other imported surface water supplies would not be enough to satiate California users, whether domestic, industrial, or agricultural, and that pumping would continue: "The problem is that today there have been more contracts signed for water in Southern California than there is water available. . . . If the California Aqueduct can't deliver, pumping will begin again . . . [and] more pumping will, of course, result in more subsidence" (Becker, 1981a; 1981b).

CONCLUSION

With the 1977 land subsidence photograph, Poland and the USGS brought subsidence science and the consequences of disappearing groundwater to a wider audience, both within and beyond the scientific community. And, with his scientific method, his desire to provide wide education and leadership, and this specific photographic product, Poland was and has remained a stalwart contributor to developing public discourse on land subsidence.

Since the 1990s, groundwater scientists have replicated Poland's photographic model, appropriating his pose by the pole to offer updates on land subsidence, from Sacramento's USGS sites in the Central Valley to locations around the world, including Japan, Taiwan, and Mexico.[9] Since its creation in 1977, the photograph has served as a visual set-point in a global focus on groundwater and land subsidence in water security science, policy, urban development, and popular environmental politics education. The photograph has become *the* standard image representing the concept of land subsidence, what many professionals and students of hydraulic systems call, in a colloquial shorthand, "that subsidence picture of the man standing next to the pole." The photograph is a graphic sound bite to any audience, from hydraulic scientists to politicians and the lay public. While the idea the image conveys is familiar—the photo has an immediate impact as transmitter of knowledge—few viewers can identify specific details: the man in the picture, the historical situation he represents, or the science behind the graphic results he displays. Even today, when I mention the photograph to water experts in any field (except groundwater study), most immediately "know" the picture, but few can identify Poland, or even realize they should.

In recent years, the urgency of the groundwater crisis has resurfaced as state groundwater management policy revives a host of historical environmental concerns—bringing Poland's photograph with it. Related crises include not only groundwater and oil pumping, but also subsurface geologic change and water contamination from subterranean mining, gas extraction, fracking, industrial runoff, hydrocompaction through extreme surface water seepage, natural peat removal or sublimation, and infrastructure collapse. Broadly visually accessible, the photo has pried open conversation onto more than the science of land subsidence; it also carries an emotional and ethical punch, a sign that signifies an unavoidable new "knowing"—that land subsidence signifies the near-permanent depletion of an essential groundwater supply we cannot do without. The photographic image opens critical questions onto a host of scientific and cultural concepts at once, questions that lead to action through exposure and discussion of the once-invisible problem. And, Poland's entry into Central Valley hydrogeology in the 1950s, grounded in his maverick experience in other parts of the state, made him a contemporary with the period of time when land and water science developments paralleled the social justice movements that accompanied them and the social documentary practice that recorded and popularized them.

By the time of his retirement in the 1970s, Poland characterized the groundwater crisis in California's Central Valley as "probably the most severe case of subsidence in the world" (Barnes n.d.). In 1999, the subsidence in the San Joaquin Valley was considered the largest human alteration of the Earth's surface (Galloway and Riley 1999). Poland's answer to the water-overuse crisis, from as early as his initial work in the 1930s and extending throughout his career and into the 1980s, had consistently been that only "conservation of water, particularly in agricultural use," could halt the conditions that led to land subsidence. Even then, he added: "It's a political problem" (Becker, 1981b; "Water Conservation Urged . . ." c. 1940s–1950s).

This brings us to now. In 2014, California gubernatorial policy crafted the Sustainable Groundwater Management Act (SGMA), a plan for statewide water use accountability and groundwater stewardship into and beyond the middle of this century. The policy was built upon groundwater management political foundations from the 1990s that did not get fully off the ground. By 2019, water districts had met a first-stage SGMA requirement to convene and submit groundwater management plans to the state water commission, with subsequent phases of implementation requirements on the books into 2042, "the date by which groundwater basins must achieve their sustainability goal" (NGO Groundwater Collaborative n.d.). The effects of this regional policy reach deep into water futures in California, and around the world, as California's plan serves as both model and companion to global frameworks like the United Nations framework Strang discusses in this volume, and presages a future consciousness of water's multiplicities, such as Klaver suggests. Finally, such grand-scale political and cultural actions are actively engaging with the appeal for long-term effects that scientific and social documentary photographers have made through their images of the land and people of the Central Valley's surface, proactive work designed to expose, and ultimately to rectify, the pressing yet invisible water problems beneath.

ACKNOWLEDGMENTS

Thank you to the University of California, Merced Center for the Humanities Water Seminar Postdoctoral Fellowship (2015–17) for generous research, staff and faculty support. Gratitude goes to Michelle Sneed, hydrologist and land subsidence specialist at the USGS California Water Science Center and UNESCO Land Subsidence International Initiative, who in 2015 offered the opportunity to research, organize, and catalogue Joseph Poland's papers. To Dan Nelson, former executive director of the San Luis & Delta-Mendota Water Authority, for sharing his time and his knowledge of California Central Valley water. To Abigail E. Owen, assistant teaching professor, Department of History, Carnegie-Mellon

University, for a shared research week in 2015 investigating groundwater science at the USGS California Water Science Center and in Central Valley field work. I thank the authors in this volume for their dedicated interest, with special thanks to my coeditors, Kim De Wolff and Ignacio López-Calvo. I dedicate this work to my parents, extending deepest thanks to them for making California water my thing.

REFERENCES

"All Water Wells Pegged." 1947. Newspaper clipping, Joseph Poland Papers, USGS California Water Science Center.

Barnes, Paul. N.d. "Like Venice, San Joaquin Valley Is Sinking." *Sacramento Union*, Metro Today Section, 1–2.

Becker, Terry. 1981a. "Commencement = Achievement: They Call Him 'Mr. Subsidence,' and Now This Old Pro Earns His Ph.D. at Age 73." *Stanford Observer*, c. June 14, 1981, 4.

———. 1981b. "Joe Poland Found Valley Was Sinking in 'Thirties; He's Back for Ph.D. at 73." *Campus Report*, June 3, 1981, 23.

Benjamin, Walter. (1931) 1977. "A Short History of Photography." Translated by Phil Patton. *Artforum*, February 1977. https://www.artforum.com/print/197702/walter-benjamin-s-short-history-of-photography-36010.

———. (1936) 1968. "The Work of Art in the Age of Mechanical Reproduction." In *Illuminations*, edited by Hannah Arendt, 214–18. London: Fontana.

Bentmann, Reinhard, and Michael Müller. 1992. *The Villa as Hegemonic Architecture*. Translated by Tim Spence and David Craven. New Jersey: Humanities Press.

Borchers, James W., ed. 1998. *Land Subsidence Case Studies and Current Research: Proceedings of the Dr. Joseph F. Poland Symposium on Land Subsidence*. Association of Engineering Geologists; Belmont, CA: Star Publishing Company.

Borchers, James W., and Michael Carpenter. 2014. *Land Subsidence from Groundwater Use in California: Full Report of Findings*. Sacramento: California Water Foundation.

Cosgrove, Denis E. 1998. *Social Formation of Symbolic Landscape*. Madison: University of Wisconsin Press.

Cotter, Jennifer. 2016. "New Materialism and the Labor Theory of Value." *the minnesota review*: 87 (1): 171–81.

Curtis, David J., Nick Reid, and Guy Ballard. 2012. "Communicating Ecology through Art: What Scientists Think." *Ecology and Society* 17 (2): 3. www.jstor.org/stable/26269030.

Daniels, Stephen, and Denis Cosgrove. 1988. "Introduction: Iconography and Landscape." In *The Iconography of Landscape: Essays on the Symbolic Representation, Design and Use of Past Environments*, edited by Denis Cosgrove and Stephen Daniels. Cambridge: Cambridge University Press.

Eco, Umberto. 1989. *The Open Work*. Cambridge, MA: Harvard University Press.

Finnegan, Cara A. 2003. *Picturing Poverty: Print Culture and FSA Photographs*. Washington, DC: The Smithsonian Institution.

Frank, Dana. 1998. "White Working-Class Women and the Race Question." 54: 80–102.

Galloway, Devin, and F.S. Riley. 1999. "San Joaquin Valley: California Largest Human Alteration of the Earth's Surface." In *Land Subsidence in the United States*, U.S. Geological Survey Circular, 12–34. Reston, VA: U.S. Department of the Interior.

Gersdorf, Catrin. 2004. "History, Technology, Ecology: Conceptualizing the Cultural Function of Landscape." *Icon* (International Committee for the History of Technology [ICOHTEC]) 10: 34–52.

Gesner, Charles. 1941. "Contamination of Water by Sea Threatens Large County Area." *Daily News, Los Angeles,* May 31, 1941.

Goodwin, James. 1998. "The Depression Era in Black and White: Four American Photo-Texts." *Criticism* 40 (2): 273–307.

Gordon, Linda. 2006. "Dorothea Lange: The Photographer as Agricultural Sociologist." *Journal of American History* 93 (3): 698–727.

Gordon, Linda, and Gary Y. Okihiro. 2006. *Impounded: Dorothea Lange and the Censored Images of Japanese American Internment.* New York: W.W. Norton.

Gunckel, Colin. 2015. "Building a Movement and Constructing Community: Photography, the United Farm Workers, and El Malcriado." *Social Justice* 42, no. 3/4 (142): 29–45.

"Heavy Water Pumping Endangers Entire West Basin: Overdraft Twice as Heavy as Before, U.S. Geologist Says in Official Report." 1946. *Inglewood Daily News,* May 21, 1946, 2. Joseph Poland Papers, USGS California Water Science Center.

Jay, Martin. 1996. *The Dialectical Imagination: A History of the Frankfurt School and the Institute of Social Research, 1923–1950.* Berkeley: University of California Press.

Kahrl, William L. 1979. *The California Water Atlas.* Sacramento: State of California.

Lange, Dorothea, and Paul S. Taylor. 1939. *An American Exodus: A Record of Human Erosion.* New York: Reynal & Hitchcock.

Latour, Bruno. 2014. "The More Manipulations, the Better." In *Representation in Scientific Practice Revisited,* edited by Catelijne Coopmans, Janet Vertesi, Michael Lynch, and Steve Woolgar, 347–50. Cambridge, MA: MIT Press.

Mitchell, Don. 1996. *The Lie of the Land: Migrant Workers and the California Landscape.* Minneapolis: University of Minnesota Press.

———. 2012. *They Saved the Crops: Labor, Landscape, and the Struggle over Industrial Farming in Bracero-Era California.* Athens: University of Georgia Press.

Mitchell, W.J.T. 2002. Introduction to *Landscape and Power,* edited by W.J.T. Mitchell, 1–4. Chicago: University of Chicago Press.

Neimanis, Astrida, Cecilia Åsberg, and Johan Hedren. 2015. "Four Problems, Four Directions for Environmental Humanities: Toward Critical Posthumanities for the Anthropocene." *Ethics and the Environment* 20 (1): 67–97.

NGO Groundwater Collaborative. N.d. *Understanding SGMA and All about the Sustainable Groundwater Act.* http://cagroundwater.org/?page_id=516 and http://cagroundwater.org/?page_id=25, accessed July 30, 2020.

Olwig, Kenneth. 1993. "Sexual Cosmology: Nation and Landscape at the Conceptual Interstices of Nature and Culture; or What Does Landscape Really Mean?" In *Landscape: Politics and Perspectives,* edited by Barbara Bender, 307–34. Oxford: Berg.

Pang, Alex Soojung-Kim. 1997. "Visual Representation and Post-Constructivist History of Science." *Historical Studies in the Physical and Biological Sciences* 28 (1): 139–71. www.jstor.org/stable/27757789.

Perini, Laura. 2006. "Visual Representation." In *The Philosophy of Science: An Encyclopedia,* edited by Sahotra Sarkar and Jessica Pfeifer, 863–70. New York: Routledge.

Perlman, David. 1967. "The Sinking Cities of the Peninsula." *San Francisco Chronicle: Voice of the West,* June 29, 1967.

————. c. 1980. "Remarks by Joe Poland (hand-edited typescript)." Joseph Poland Papers, Box 2, Folder 2.3. USGS California Water Science Center.

————. 1984. "Case History No. 9.14. Santa Clara Valley, California, U.S.A." In *Guidebook to Studies of Land Subsidence Due to Ground-Water Withdrawal: Prepared for the International Hydrological Programme, Working Group 8.4*, 279–90. Paris: UNESCO.

Poland, Joseph, and Ben E. Lofgren. 1984. "Case History No. 9.13. San Joaquin Valley, U.S.A." In *Guidebook to Studies of Land Subsidence Due to Ground-Water Withdrawal: Prepared for the International Hydrological Programme, Working Group 8.4*, edited by Joseph Poland, 263–77. Paris: UNESCO.

Poland, Joseph F., Ben E. Lofgren, Richard L. Ireland, and A.G. Pugh. 1975. *Land Subsidence in the San Joaquin Valley, California, as of 1972*. U.S. Geological Survey Professional Paper 437-H. Sacramento: USGS.

Prokopovich, Nikola P. 1973. "Engineering Geology and the Central Valley Project." *American Water Works Association Journal* 65 (3): 186–94. www.jstor.org/stable/41266925.

Puskar, Jason. 2016. "Black and White and Read All Over: Photography and the Voices of Richard Wright." *Mosaic: An Interdisciplinary Critical Journal* 49 (2): 167–83. www.jstor.org/stable/44030590.

Richards, Harvey. 1965. "Dolores Huerta HUELGA Sign—Without Doubt the Iconic Photo of the Delano Grape Strike 1965." UCSD Library. "Dolores Huerta 1965" album, Harvey Richards Archive, Farmworker Movement Online Gallery. San Diego, CA. https://libraries.ucsd.edu/farmworkermovement/gallery/thumbnails.php?album=474.

Riley, Francis S. 1998. "Mechanics of Aquifer Systems—The Scientific Legacy of Joseph Poland." In *Land Subsidence Case Studies and Current Research: Proceedings of the Dr. Joseph F. Poland Symposium on Land Subsidence*, edited by James W. Borchers, 13–27. Association of Engineering Geologists; Belmont, CA: Star Publishing Company.

Roberts, Liz. 2017. "Geographic Methods: Visual Analysis." *Oxford Bibliographies*, September 27, 2017. DOI: https://doi.org/10.1093/OBO/9780199874002-0173.

Rudwick, Martin J.S. 1976. "The Emergence of a Visual Language for Geological Science, 1760–1840." *History of Science* 14 (3): 149–95.

Sandweiss, Martha A. 2002. *Print the Legend: Photography and the American West*. New Haven: Yale University Press.

Shiff, Richard. 1992. "Handling Shocks: On the Representation of Experience in Walter Benjamin's Analogies." *Oxford Art Journal* 15 (2): 88–103. www.jstor.org/stable/1360503.

Shiffman, Dan. 2007. "Richard Wright's '12 Million Black Voices' and World War II-Era Civic Nationalism." *African American Review* 41 (3): 443–58.

Sneed, Michelle, Justin T. Brandt, and Mike Solt. 2018. *Land Subsidence along the California Aqueduct in West-Central San Joaquin Valley, California, 2003–10*. Scientific Investigations Report 2018-5144. Reston, VA: U.S. Geological Survey. https://pubs.usgs.gov/sir/2018/5144/sir20185144.pdf.

Snyder, Joel. 2002. "Chapter 6: Territorial Photography." In *Landscape and Power*, edited by W.J.T. Mitchell, 175–201. Chicago: University of Chicago Press.

Starr, Kevin. 2004. Foreword to *Photographing Farmworkers in California*, by Richard Steven Street, xi–xv. Stanford: Stanford University Press.

Street, Richard Steven. 2004. *Photographing Farmworkers in California*. Stanford: Stanford University Press.

———. 2008. *Everyone Had Cameras: Photography and Farmworkers in California, 1850–2000*. Minneapolis: University of Minnesota Press.

Taylor, Paul. c. 1935. *Dorothea Lange in Texas on the Plains, circa 1935*. Oakland Museum of California, Oakland. https://museumca.org/file/dorothea-lange-texas-plains-circa-1935-paultaylorlo-resjpg.

Tolman, C.F., and Joseph F. Poland. 1940. "Ground-Water, Salt-Water Infiltration, and Ground Surface Recession in Santa Clara Valley, Santa Clara County, California." *Transactions of the American Geophysical Union* 21 (Part 1): 23–34.

USGS. 2018. *Location of Maximum Land Subsidence in U.S.* March 8, 2018. www.usgs.gov/media/images/location-maximum-land-subsidence-us-levels-1925-and-1977.

———. 2019. *Land Subsidence in the United States, USGS Fact Sheet-165–00*. http://water.usgs.gov/ogw/pubs/fs00165/.

Veblen, Thorstein. 1899. *The Theory of the Leisure Class*. New York: Penguin Publishing Group.

"Water Conservation Urged in [Long Beach] Area." c. 1940s–1950s. Newspaper clipping, Joseph Poland Papers, USGS California Water Science Center.

Wright, Richard, and Edwin Rosskam. 1941. *12 Million Black Voices: A Folk History of the Negro in the United States*. New York: Viking Press.

NOTES

1. The benchmark (BM) Poland reference is S661, also recorded as GU0103 at the ArcGIS website: www.arcgis.com/home/webmap/viewer.html?webmap=00ea57b3f73e43d1b0ae57f937bea633&extent=-120.5493,36.6637,-120.4859,36.6919, accessed July 24, 2020. Although the physical benchmark placed before 1977 was no longer found by the 1980s, the USGS now locates the spot by its GPS coordinates. The photograph with captions appears on the USGS website: www.usgs.gov/media/images/location-maximum-land-subsidence-us-levels-1925-and-1977, accessed July 24, 2020. Map 7.1 in this chapter shows the physical evidence of subsidence between 1926 and 1970, which the placards in Poland's 1977 photograph displayed in abbreviated terms. For a map that shows older (1926–70) and newer (2008–10) subsidence, see figure 17 at https://pubs.usgs.gov/sir/2018/5144/sir20185144.pdf (Sneed, Brandt, and Solt 2018).

2. On the basic relationships between aquifers, aquitards, and water in the context of the Central Valley: "An aquifer system contains aquifers and aquitards. Aquifers are composed of larger grains (sands and gravels) that transmit water easily. Aquitards are composed of smaller grains (clays and silts) that do not transmit water easily. Both aquifers and aquitards have spaces between the grains that store water. When an aquifer system is pumped, water comes from both aquifers and aquitards. Some folks use the term 'aquifer' to include both 'aquifers and aquitards' as I've defined above. In this way, they think of the entire system as an aquifer and aquitards are part of the aquifer." Michelle Sneed (hydrologist and land subsidence specialist, California Water Science Center, USGS), in communication with the author, July 2020.

3. See figure 17 at https://pubs.usgs.gov/sir/2018/5144/sir20185144.pdf (Sneed, Brandt, and Solt 2018). For a fuller context of California water at the time of Poland's photograph, see also Kahrl (1979).

4. Poland to Dr. Kirtley F. Mather of Harvard University's Department of Geology, August 16, 1932, in Joseph Poland Papers, Box 2, Folder 2.34, "Consulting Correspondence 1930s," USGS California Water Science Center.

5. Sneed, in communication with the author, July 2020.

6. Poland to Mather, August 16, 1932.

7. It is a utility pole in Poland's photograph, but for some viewers it replicates a protruding well casing. In fact, "it is not uncommon for a well casing to protrude from the ground as a sign of land subsidence." Sneed, in communication with the author, July 2020.

8. C.F. Tolman to Joseph Poland, February 16, 1940, in Joseph Poland Papers, Box 2, Folder 2.22, USGS California Water Science Center.

9. Sneed, in communication with the author, July 2020.

Irrigated Gardens of the Indus River Basin

Toward a Cultural Model for Water Resource Management

James L. Wescoat Jr. and Abubakr Muhammad

The cultural connections among water, gardens, humanities, and policy span more than five thousand years in the Indus River basin, from the archaeobotany of Harappan floodplain settlements to the historical geography of Mughal gardens and waterworks, to critical histories of massive colonial canal irrigation and searching reflections on postcolonial waterscapes. In this chapter we ask how policy inspiration can be drawn from the composite culture of irrigated gardens in the Indus basin. This effort at the basin scale bears comparison with Victoria Strang's essay in this volume on the role of cultural values at the global scale in United Nations water policy. In between these basin and global scale perspectives lie expansive new studies of "Asia's" water resources (e.g., Amrith 2018; Chellany 2013; Ray and Maddipati 2020). While helpful for extending the boundaries of analysis to a macroregional scale, they tend to emphasize geopolitical histories of water development, particularly in the Indus, which is an important perspective that we acknowledge but seek to move beyond. The literature on cultural values of water in the Indus, by comparison, includes historical relationships among the arts, religion, caste, honor, gender, and tribal as well as territorial politics (e.g., Gilmartin 2015; Meadows and Meadows 1999; Mustafa 2013; Naqvi 2013). Scholars from all disciplines in the humanities have contributed to this emerging perspective, especially historians, albeit in small numbers and with limited reception. It seems fair to say that cultural research has contributed more to policy history and critique and less toward policy alternatives of the sort that Strang suggests and that this chapter seeks to imagine.

Among the myriad coexisting, often conflicting, cultural values of water at work in the Indus, we focus here on the generative metaphor of "irrigated gardens" (Schön 1979). We show how irrigated gardens have developed in inspiring ways from the Indus headwaters in the Hindu Kush and Karakorum mountains to the plains of the middle Indus basin, and ultimately out into the Indus delta and Arabian Sea. Our approach is not strictly limited by Indus River basin boundaries, for its relevance extends to adjoining landscapes that drain into the Amu Darya, Helmand, Yamuna, Ganga, and Luni rivers. We concentrate on the Indus basin as a case study, however, to envision a new approach to its management and care. Our emphasis on garden ideals thus has normative and historicist, as well as historical, aims. We argue for what could be, and what could be better, if the jointly humanistic and scientific ideals of irrigated gardens were adopted in water resources research and management.

A CULTURAL PERSPECTIVE ON THE IDEA OF THE INDUS AS AN IRRIGATED RIVER BASIN

The Indus River basin is one of the world's international water resources and irrigation laboratories (map 8.1). It has been the focus for massive programs of physical transformation, investment, and research on irrigation and drainage systems (Wescoat, Halvorson, and Mustafa 2000; Yu et al. 2013). Before evaluating that record, it is interesting to ask: how did the Indus come to be regarded as a basin? Any search for origins is elusive. At the deepest historical level, one may speculate that riparian settlements of the protohistoric Indus Valley civilization were part of a common socio-hydrologic region during the fourth to second millennia BCE. Harappa itself had richly irrigated plantings of cereals, millets, oil seeds, vegetables, and fruits (Weber 1999). However, Harappan cultural sites were not limited to the Indus valley. They extended into the Yamuna basin, the Thar Desert, and the Saurashtra coastal region, well beyond the Indus (Wright 2010).

The Indus River channel was known as Sind in antiquity. Alexander of Macedon famously crossed the Sind in 327 BC before his final battle against a regional ruler named Porus, who is known throughout Punjab to this day (McCrindle 1896). Ptolemy's second-century-BCE *Geography* referred to the river as Sind, from the Sanskrit place-name *Sindhu*, and the river was inscribed as such on countless maps from the thirteenth to the sixteenth century, when the postclassical Latin word Indus was adopted. Timurid histories of this period described the origin of the Jhelum tributary in Kashmir, its confluence with the Indus main stem, and ultimate discharge into the Arabian Sea (e.g. Yazdi 1976, 521–52). A Mughal map copied in the eighteenth century clearly delineates the Indus main stem and five major tributaries of the Punjab ("five waters") region, which began to convey the

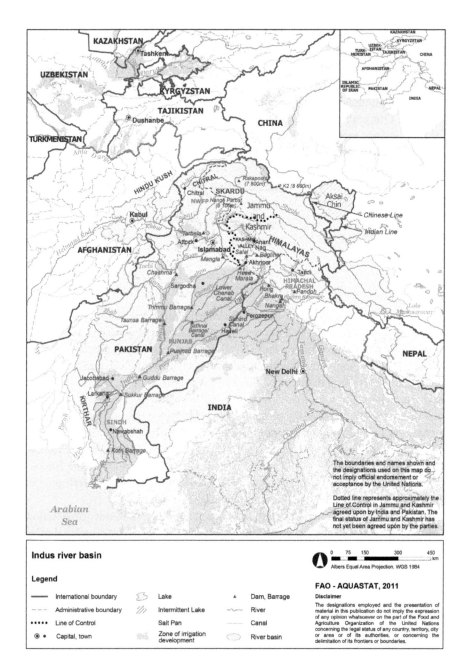

MAP 8.1. INDUS River Basin map. UN FAO, Aquastat, 2011, Indus River Basin map, url: https://storage.googleapis.com/fao-aquastat.appspot.com/PDF/MAPS/indus-map_detailed .pdf). Reproduced with permission.

broader sense of a tributary river network (Phillimore 1952). Some painted maps and silk tapestries of this period also depicted large water bodies and associated plantings in the region (Gole 1989, 116–32).

Even though scientific delineation of *bassins* only developed with the work of French geographer Philippe Buache in the early eighteenth century, most nineteenth-century accounts continued to treat the Indus, and other rivers, as channels and not as basins (Buache 1757; Wescoat 2017). The term *Indus basin* was used in *A Catalogue of the Plants of the Punjab and Sindh* by the botanist James Edward Tierney Aitchison (1869) and in late-nineteenth-century geographical descriptions of the region by Clements Markham, Elisee Reclus, and others. One also finds geological references to the "upper Indus" as a basin in the same period (Drew 1874; Greenwood 1874).

British irrigation works were sometimes organized by basin, omitting the word *river*, for example, Ganges basin, Indus basin, and so on. However, the major irrigated provinces of the Indus basin—Punjab and Sindh—developed separately from and often in conflict with one another over basin water resources from the advent of British canal irrigation projects in the 1850s onwards (Gilmartin 2015).

The river basin concept was central to David E. Lilienthal's (1951) argument for treating the Indus as a shared context for water resource development in independent India and Pakistan. Eight years of negotiation culminated in the Indus Waters Treaty of 1960, along with the Indus Basin Development Fund created by the World Bank in 1961 to implement the infrastructure investment provisions of the treaty, and it was in the latter context that the term *Indus basin* came into widespread use. Interestingly, the Indus Waters Treaty refers twice to "household gardens and public recreational gardens," excluding these domestic and civic uses from regulation.

Thus, the concept of the Indus River basin may have had ancient origins, but it has had only recent usage for water resources management purposes since the mid-twentieth century. While the expression *Indus basin* implies a geographically shared natural resource, the Indus Waters Treaty of 1960 sacrificed that ideal for a division of the tributary river channels between India and Pakistan (see the extensive treatments in Adeel and Wirsing 2017; Amrith 2018; Gulhati 1973; Haines 2016; Hussain 2017; Michel 1967).

CULTURAL INTERPRETATIONS OF THE INDUS RIVER BASIN AS A WATER SYSTEM AND FOOD MACHINE

The Indus Waters Treaty of 1960 coincided with three major advances in river basin planning and development. Financially, to support the treaty that it helped broker, the World Bank created an Indus Basin Development Fund with support from a consortium of "friendly countries" that funded investments in dams, link canals, and other major infrastructure works (on a scale analogous to the Midi canal in Mukerji's chapter in this volume). New models and analytic methods helped guide these investments. They built upon advances in computing power to generate synthetic hydrologic flow series, develop hydro-economic optimization models, and evaluate development alternatives (Maass et al. 1962). Applications of systems analysis in the water resources field are attributed in part to the Harvard Water Program. The system in this case involved a network of channels to deliver water to different agro-economic zones (Ahmad and Kutcher 1992; Duloy and O'Mara 1984).

The systems approach is itself a cultural model. Systems analysis emerged out of the field of cybernetics at a time of transdisciplinary mathematical and computational research that introduced systems thinking into water resources research (Wiener 1961). Simulation modeling identifies components, links them, and then alters those components in various ways to assess their effects. An optimization model, by comparison, searches for the best solution to any change in conditions. The World Bank's Indus Basin Model optimizes the net economic benefits of infrastructure investment and policy reform alternatives (Ahmad and Kutcher 1990; Yu et al. 2013). This idealized vision of optimal solutions to complex water problems captivated the collective engineering imagination during the second half of the twentieth century, and it was some time before its elusive qualities were realized and addressed with modified strategies of seeking solutions deemed good, or good enough, by society and policy makers.

The water engineering culture has been accompanied by an institutional culture evolving to address transboundary water issues (Michel 1967). It is interesting to note that the Harvard Water Project was led by political scientist Arthur Maass (1962). An institutional challenge at the international level is that India and Pakistan have federal systems of government, in which water is primarily a state subject with intense interstate water conflicts, which clashes with the strong centralized water organizations created after independence. For example, inspired in part by the Tennessee Valley Authority, Pakistan created a central Water and Power Development Authority (WAPDA) for a basin five times the size of the Tennessee River basin!

In a third development following independence, the Indus basin came to be conceived as a "food machine," which is once again an Anglo-American cultural construct (see Wescoat [2013] on its late-eighteenth-century origins). Professor Roger Revelle (1985) seems to have coined the food machine metaphor during his team's research on waterlogging and salinity of irrigated lands in Pakistan, commissioned by President John F. Kennedy during the regime of President Ayub Khan in the early 1960s. This was the era of the Green Revolution, when India and Pakistan were not yet food secure, and the food machine metaphor implied a visionary remedy for the problem, which critics later cast as socially and ecologically damaging (Johnson, Early, and Lowdermilk 1977; Kango 1997). Even so, the machine metaphor continues to be applied or evoked in large river basins of South Asia (e.g., Acciavati 2015).

THE SHIFT TO WATER RESOURCES MANAGEMENT:
A CULTURAL INTERPRETATION

While the engineering systems approach and machine metaphor proved fruitful for physical infrastructure development, their outcomes on the ground posed jointly material and cultural issues. These included the environmental and social impacts of dams, waterlogging and salinity from canal seepage, groundwater depletion as millions of private tubewells supplemented surface water deliveries, and a rural-to-urban shift in water demand (Wescoat, Halvorson, and Mustafa 2000). Widespread social inequities in water deliveries, especially in the tail end of canals, contributed to lower crop yields than in other regions (Briscoe and Qamar 2005; Mustafa 2013).

"Management" became the cultural and policy innovation for addressing these problems. The shift from infrastructure to management encompassed multiple scales and sectors. At the local level, gaps between infrastructure investment and agricultural yields were attributed to poor water management by farmers at the field level. Management reformers sought to link improved canal deliveries by provincial irrigation departments to on-farm support for farmers administered by provincial departments of agriculture (World Bank 2017). The International Irrigation Management Institute (later the International Water Management Institute, see IWMI n.d.), created in 1984, brought together an expanding network of anthropologists, geographers, and sociologists who drew attention to cultural as well as operational issues (e.g., Merrey 1986).

At the highest level, the top-down approach described in Veronica Strang's chapter in this volume, the Global Water Partnership advanced a comprehensive approach called Integrated Water Resources Management (IWRM). The IWRM approach combined earlier models of integrated river basin development with increased emphasis on water resource economics, institutions, environment, and gender issues (Lenton and Muller 2009). It too faces criticism for eliding the political and cultural roots of mismanagement, economic distortion, hazards, and social injustice (Mehta, Derman, and Manzungu 2016; Mollinga, Athukorala, and Dixit 2006). For example, East Punjab the celebrated "breadbasket" of India in the 1960s now faces deepening socioeconomic problems that include groundwater depletion and farmer suicides (Singh, Bhangoo, and Sharma 2018), while Kashmir has suffered violent multidecadal conflict (Rao 2008).

There have been innovative efforts to address these weaknesses in the management paradigm. Elinor Ostrom (1990; 1993) developed a framework of Institutional Analysis and Design (IAD), which emphasized "rules in use," that is the actual social relationships, modes of communication, and informal rules that govern social action, rather than formal written policies that obscure as much as they explain. Water governance is a rapidly expanding focus of research. The emerging field of socio-hydrology strives to couple hydrologic and social processes through historical, comparative, and analytical methods (Wescoat 2013; Wescoat, Muhammad, and Siddiqi 2018). These approaches have made important contributions, but they have met with limited success in turning around troubled irrigation systems. *It seems clear that something is missing and that a fresh cultural approach is needed.*

"IRRIGATED GARDENS" OF THE INDUS RIVER BASIN

We propose to think about river basins as complex historical and cultural processes in which the humanities play a vital role (Wescoat 2012; Muhammad 2016). The humanities have been largely implicit or limited in the approaches to water management discussed above, even though there is an abundance of relevant and available material, which includes histories of monsoon rainfall, irrigated landscapes, gardens, water architecture, and more (Rajamani, Pernau, and Schofield 2018). The humanities have explored water folklore, water aesthetics, water symbolism, water rituals, and still more (see Meadows and Meadows [1999] and Naqvi [2013] for Indus examples). We call this perspective "cultural" in a renewed sense of the Berkeley school of cultural geography, which has focused on environment-society relationships in agrarian landscapes (e.g., Leighly [1963] and Wallach [1996], and critiques by Agarwal and Sivaramakrishnan [2000] and Mitchell [1996], as discussed in Faletti's chapter in this volume).

Irrigation is by far the largest use of water in the Indus and other arid river basins (Wescoat 2000). To probe the humanistic dimensions of these irrigated landscapes, we undertake a geographical study of irrigated gardens that provide promising models of creative water management from the headwaters to the delta.

Gardens exist in many forms, some metaphorical and others literal. Even Roger Revelle (1985, 32), who characterized the Indus basin as a food machine, wrote, "I think that the main thing we did, in reality, as opposed to reputation or talk, was our insistence that this could be a Garden of Eden, that this was a tremendous agricultural resource and could be developed in a very profitable, very useful [way] for the people of Pakistan."

There are some pitfalls to be avoided with the irrigated garden concept. Take the famous Mughal gardens of Lahore, in which the ideals of paradise gardens in the Qur'an contrast starkly with the historical realities of those gardens (Wescoat 1991; 1995). Mughal gardens had the physical characteristics ascribed to paradise gardens but lacked the Qur'anic ideals of religious faith and good works. Late-twentieth-century research has drawn attention to such gardens as spaces of territorial domination and social control (Mitchell 2002). The Qur'an has several verses that warn about the destruction of worldly gardens that produced thistles rather than fruit, and were destroyed for the hubris they embodied (e.g., Qur'an 18:32–43; 44:25).

There are other places that were not called gardens but that have had some of the spiritual qualities of paradise that we seek in the Indus basin. We envision the basin as a mosaic of such gardens, including small horticultural plots and urban gardens, which have influenced one another upstream and down. Before surveying this mosaic, we briefly consider the history of ideas about gardens in the region.

The earliest evidence of gardens comes from ethnobotanical centers of Harappan civilization (third to second millennia BCE). As noted above, Harappa supported horticultural crops of fruits and vegetables that included melons, dates, and grapes as well as grains (Weber 1999; Weber and Belcher 2003). While the ideological character of Harappan cultivation is unclear, horticulture was foundational to this earliest civilization in the Indus basin (Wright 2010).

By comparison, Sanskrit literature of the first millennium CE associated gardens with groves, orchards, and parks (*udyana, apavana, sada-pushpa, vata*, etc.). Some major themes in those gardens involved pleasure, play, and romantic trysts in forest landscapes; with flowers, fruits, and kitchen gardens in domestic enclosures (Apte 1957–59; MacDonnell 1929). The celestial gardens of Indra, Krishna, and the *devas* were said to be located on the eastern side of the sacred Mount Meru near two large pools. References to gardens in Pali literature (e.g. *amba, arama, iddhi, uyyana*) likewise stress pleasure gardens associated with parks, groves, and natural streams (Pali Text Society 1921–25).

Modern Hindi and Urdu concepts of gardens have even broader connotations (Platts 1884). Some derive from Sanskrit (e.g., *upavan*, which refers to planted groves; and *udayan*, to royal gardens). Other words build upon the early Persian garden concept of the *bagh*, which carried over into the regional languages of Punjabi, Pushto, Sindhi, and Urdu. These modern languages stress practices of gardening, as in orchards, kitchen gardens, fields, and plantations. *Bagh* remains a common word for gardens in South Asia and is variously associated with the *chahar bagh* (fourfold garden) and *baghicha* (small gardens). The related Persian word *bustan* variously denotes orchards, flower gardens, and fragrance gardens. Other Urdu-Persianate garden terms include *pushpa* and *phulwari* (flower garden), *gul* and *gulistan* (rose garden), *jannat* and *firdaws* (paradise), *rauza* (tomb garden), *chamman* (meadow), *kiari* (planting beds), and more. Gardening castes and tribal groups of the north Indian region included *Arains, Baghbans, Malis*, and *Kachwaris* (Ibbetson and Maclagan 1911–19).

With these historical and linguistic notes in mind, we move to the search for irrigated garden precedents in the Indus, from its headwaters downstream to the delta. This downstream approach stands in contrast with upstream narratives that struggle against the current and terrain, in ways that have been associated with pilgrimage journeys and colonial conquest (Wescoat 2018). The sequence of subregions considered is:

1. Upper Indus River above Tarbela Dam;
2. Kabul River and its tributaries above the Indus main stem;
3. Upper Jhelum and Chenab basins of Kashmir and Jammu;
4. Punjab tributaries (Hills, East and West); and
5. Lower Indus basin from Panjnad to the delta.

When finished, we reappraise the implications of this composite mosaic of irrigated garden cultures in the Indus basin.

1. Irrigated Gardens of the Upper Indus River Basin

Irrigated gardens of the Upper Indus exemplify the garden ethos espoused in this study. The Upper Indus basin is bounded by the Himalaya, Hindu Kush, and Karakorum mountain ranges, sometimes called the water towers of Asia, that give rise to a dynamic regime of snow and ice hydrology (Bianca 2005; ICIMOD 2018; Khan et al. 2014; Kreutzmann 2006). Major subregions include Chitral, in the far northwest; Gilgit and Baltistan, which flank the upper Indus mainstem and its Shyok tributary; and Ladakh, which drains the northeastern headwaters of the Indus on the margins of China and Tibet.

Notwithstanding its high-altitude remoteness (Hussain 2015), the upper basin has a rich heritage of irrigated gardens that range from orchards to terraced fields and countless household kitchen gardens (*basi* in Burushaski and Shina languages). An extensive *Grammar of Shina Language and Vocabulary* features 146 entries for "Farming, gardening, trees, vegetables, fruits, and flowers" (Rajapurohit 2012, 102–7). Irrigation channels cut from small tributaries run by gravity for kilometers along the contours of steep hillslopes.

An important collection of essays drew attention to traditions of sharing water that are central to the garden ideals introduced earlier (Kreutzmann 2000). These villages celebrate irrigation festivals that build solidarity through the growing season and observe water use taboos to minimize pollution and scarcity. Innovative irrigation communities construct "artificial glaciers," in which early snowmelt on south-facing slopes is diverted to north-facing slopes, where it refreezes (Angchok and Singh 2006). When this frozen water remelts later in the season, it is used downslope or diverted back around to south-facing slopes for much needed late-season irrigation water. This ingenious irrigation method entails a high level of community collaboration on multiple scales.

The Aga Khan Rural Support Programme employs traditional and modern cultural approaches to improving irrigated farms in the upper Indus basin through a combination of sociocultural mobilization, technical support, women's education, ethno-religious pluralism, and financial planning (Wood, Malik, and Sagheer 2006, 454–90). The Aga Khan Trust for Culture has a parallel program of cultural heritage conservation projects in the Karakorum region that include links between water and gardens (Bianca 2005).

Irrigated orchards and kitchen gardens make significant contributions to agrarian landscapes of the northern areas (Kreutzmann 2006; Parveen et al. 2015). Local irrigation systems have a combination of formal organizations and informal rules for water allocation by turns, canal maintenance, dispute resolution, and emergency repair. Recently, fifty-seven local gardeners formed a Chitral Horticultural Society that featured ten of its members' gardens on its website, along with plant lists and Khowar garden poetry. In addition to these physical gardens, regional folktales speak of magical gardens watered by running streams that have trees of life and death, tended by elderly couples, inhabited by fairies (*pari*), and full of moral lessons (Swynnerton 1987).

These irrigated gardens of the upper Indus combine beauty, economy, ecology, morality, and spiritual life in integrated ways. They encounter numerous hazards and display remarkable cultural adaptation and resilience. They also face challenges posed by government proposals to build massive hydropower projects on the Indus main stem. Before proceeding to those downstream connections, we need to consider the adjacent headwaters of the river Indus.

2. Irrigated Gardens of the Kabul River Basin

The Kabul River was not included in the Indus Waters Treaty, but it has a fascinating historical geography of irrigated landscapes from antiquity to the present that is relevant for the Indus. Kabul was associated with a beautiful paradisiacal city mentioned in *The Rigveda* (Jamison and Brereton 2017). Ancient Gandharan sources mention charitable wells (Falk 2009) associated with monasteries and settlements north of the Kabul River and in the tributary Swat River valley.

Some of the early irrigated gardens and waterworks of Kabul were built by the founder of the Mughal dynasty, Zahiruddin Muhammad Babur, from 1504 to 1526 CE (Parodi 2021, forthcoming; Wescoat 1991). A famous image of Babur laying out the *Bagh-i Wafa* (Garden of Fidelity) downstream from Kabul near Jalalabad draws together formal, symbolic, and experiential garden qualities (Wescoat 1989, 2021). The *Baburnama* makes it clear that gardens were places of cultivation, poetry, camaraderie, and sensory enjoyment. While many of these sites have been lost, Babur's terraced funerary garden of fruit trees along a central water axis, known as the *Baghe Babur*, has been recently restored by the Aga Khan Trust for Culture in Kabul. Later Mughal governors built irrigated gardens in downstream reaches of the Kabul basin in Peshawar, Naushehra, Vallai, and Swat (Rehman 1996).

The garden heritage in this region faces serious challenges, however, including decades of conflict in Afghanistan that have spilled into the Swat valley and other areas of Pakistan. Anticipating a lessening of those conflicts, proposals have recently been put forward to build a cascade of dams and reservoirs along the upper Kabul River in Afghanistan. These projects promise upstream water, agriculture, and power benefits, accompanied by downstream water and environmental costs (World Bank 2010). Similar dam proposals and controversies roil downstream in Pakistan at sites like Kalabagh, which means "black garden," which may refer to mango orchards whose dark green foliage appears black from a distance. The current situation in the Kabul River basin demands a new approach that takes upstream and downstream landscapes seriously.

3. Irrigated Gardens of the Upper Jhelum and Chenab Rivers in Kashmir and Jammu

Few places on earth have as strong an association with gardens, water bodies, and paradise symbolism as the greater Kashmir region. Few places have as great a need for a transformative approach to water and environmental management. Flooding, conflict, and controversial water development proposals afflict the region

(Rao 2008). At the same time, historical conceptions of Kashmir as a paradisiacal landscape have deep roots that can inspire a fresh cultural approach to water management (e.g., Inden 2008; Petruccioli 1985). Kashmir and neighboring Azad Kashmir drain the upper Jhelum river basin, which has a mix of snow and ice melt, while the adjacent district of Jammu drains the Chenab River, which has a combination of monsoon rainfall and snowmelt. Each of these headwaters regions faces distinct challenges and opportunities.

Kashmir has the most renowned garden history in the Indus basin and in South Asia at large. Its origin myth involves the draining of a great lake that was named *Satisar* by a powerful religious saint *(rishi)*. Water and garden verses weave throughout the poet Kalhana's epic *Rajatarangini* (River of Kings), written in the thirteenth century, which narrates the vicissitudes of good and bad rulers (Pandit 1935; Kaul 2018; Stein 1900; Zutshi 2014). The text speaks of thousands of gardens and water bodies that had widely varying qualities of beauty and danger, growth and decline, poverty and prosperity (Stein 1900).

Four centuries later, Mughal rulers built pleasure gardens along the shores of Dal Lake, and countless poems and texts proclaimed Kashmir's affinity with the gardens of paradise. When the Mughals rulers invaded Kashmir toward the end of the sixteenth century, it reminded them of a Central Asian paradise lost, as well as a new paradise found. They appropriated Kashmir's mytho-poetic paradise heritage, which they fused with their own tradition of garden building and literature (Sharma 2017). They represented water gardens in paintings. The fourth Mughal ruler Jahangir (1999) went so far as to speak of Kashmir as a regional garden, as his father Akbar had for all of Hindustan (Koch 2008). Later rulers commissioned large silk and cotton maps of this Dal Lake and River Jhelum landscape.

Today, the Indian National Trust for Art and Cultural Heritage (INTACH) features gardens and waterworks in its heritage conservation projects. Kashmiri garden ideals find expression in vernacular as well as elite courtly texts. Vegetable gardeners on the margins of Dal Lake ingeniously expanded their planting beds and yields by creating floating gardens. Floating on light wooden frames, the lake gardens have layers of rich silt and organic matter in which vegetables root and grow (Casimir 2021). Water gardens extend from these floating vegetable plots to the formal water and *chinar* (Oriental plane) tree axes in Mughal gardens to fields irrigated by mountain springs (INTACH 2014). Conservation of Kashmir's water garden heritage draws together popular, poetic, and provincial aspirations (figure 8.1).

As noted above, the Indus Waters Treaty of 1960 excludes floating domestic gardens of the sort that abound in Kashmir. Indus water developers focus instead on large hydropower projects like India's Kishanganga Dam on the Jhelum River, which competes with Pakistan's nearby Neelum-Jhelum project. International arbitration drew attention to environmental impacts downstream that neither country seemed well prepared to address, but that would be central in a garden approach.

FIGURE 8.1. Mughal water gardens in Nishat Bagh, Kashmir. Photo courtesy James L. Wescoat Jr.

There is a need to comprehensively rethink the waters of Kashmir, including their linkages with irrigated garden heritage downstream in Punjab. That vision seems remote at present, as conflict and mismanagement have wracked the region for decades, leading some to regard Kashmir as a "paradise lost" (Rao 2008; Sharma 2017). However, the region has a unique heritage and prospects for renewal as an irrigated garden.

The Jammu region in the upper Chenab River basin lies just east of Kashmir. It has a very different origin myth in which King Jambu Lochan saw a lion and a goat drinking side by side from a stream, which led him to select that peaceful site for his capital in Jammu. His brother Bahu Lochan built a garden fortress and temple on the banks of the Tawi River known as *Bagh-e Bahu* (Garden of Bahu). The Tawi and Chenab basins have historically supported irrigated rice in summer and wheat in winter, along with vegetables and medicinal plants, but they too face pressures to develop large hydropower projects.

4. Irrigated Gardens of the Punjab ("Five Rivers")

Downstream, Punjab presents new patterns of irrigated gardening that have had global as well as regional significance. Punjab comprises broad alluvial plains, which are watered in part by runoff from the former Punjab hill states in the northeastern headwaters of the basin (Malhotra and Mir 2012). For most of its

FIGURE 8.2. Irrigated orchards and fields in Punjab. Photo courtesy James L. Wescoat Jr.

premodern history, the region had extensive pastoral lands and localized areas of flood farming, inundation canals, and shallow dug wells, which supported riparian settlements and dispersed hamlets on upland *bars*, from antiquity to the early modern era (Gilmartin 2015; Mughal 1997).

These early irrigated landscapes underwent three major transformations during the Mughal, colonial, and postcolonial periods (figure 8.2). For example, Punjab has some of the earliest Mughal period gardens on the subcontinent. They include Kallar Kahar, where Babur occupied a rock outcrop carved into a throne (*takht*) that provided views of a lake and surrounding orchards. Mughal rulers also built riverfront gardens like Kamran's *baradari* along the Ravi riverfront, which emulated the riverfront gardens of other Mughal cities like Agra and Delhi. Over time, Mughal gardens became monumental in size, design, and waterworks; and the Punjab capital city at Lahore came to be known as the Mughal City of Gardens. Emperor Shah Jahan (r. 1627–57) ordered construction of a large canal to irrigate Shalamar Bagh and adjacent garden plots along the Ravi River terrace near Lahore. However, the French traveler Francois Bernier wrote a cautionary letter to Jean-Baptiste Colbert, the finance minister for Louis XIV (discussed in Mukerji's chapter in this volume), on the deterioration of water infrastructure in regimes that neglect the rights of cultivators (Wescoat 2000).

These Mughal gardens were places of dynastic pleasure and power that had the form and elements of paradise in Islam, but often lacked its ethos. For example, Kamran's baradari in early-sixteenth-century Lahore witnessed a series of internal betrayals among Mughal brothers, fathers, and sons (Wescoat 1995). Mughal paintings of gardens and waterworks provided idealized representations of what gardens ought to be as, for example, when they depicted dignified gardeners tending the soil and plantings, or Sufi mystics conversing in idyllic garden settings.

Eastern Punjab had Mughal gardens like Pinjore, and charitable drinking water wells at caravanserais along the Grand Trunk Road from Bengal to Kabul. These facilities were followed by important Sikh gardens and sacred water bodies like the Amritsarovar tank surrounding the Golden Temple (Parihar 1989; CRCI 2009). When one considers Sikh faith and practice, as we have for Islamic gardens and water rights in western Punjab (Wescoat 1995), a rich perspective emerges. The *Guru Granth Sahib* has scores of references to gardens (*baga*) and waterworks. A search for the keyword "garden" yields verses that associate the idea of the garden with God's creation: "This world is a garden, and my Lord God is the Gardener" (Sri Granth n.d., Guru Amar Das 118, line 7). The relationship between gardens and irrigation is profoundly spiritual: "The Word is the tree; the garden of the heart is the farm; tend it, and irrigate it with the Lord's Love" (Sri Granth n.d., Guru Nanak Dev 254, line 17). Sikh texts include fascinating historical garden stories as well, like the refuge sought by Guru Gobind Singh in the village garden of Macchiwara in 1704 CE. Verses on water and waterworks frequently mention pools (*pula*), a logical image given the reliance on rainfall, floodplains, and wells in precolonial Punjab (Wahi 2014).

Proceeding upstream a bit, the headwaters of the former Punjab Hill States offer additional perspectives on montane waterworks, gardens, and landscapes in the region. These small kingdoms were established in the foothills of the Himalayas from Jammu in the northwest to Garhwal in the southeast. In between lay dozens of small principalities in the upper Ravi, Beas, and Sutlej river tributaries of the Indus. We gain special insights into their irrigated gardens from exquisite local schools of painting that flourished in the late eighteenth through mid-nineteenth centuries (Archer 1973; Losty 2017). Seven examples stand out for their water and landscape imagery.

Basohli state was situated on the right bank of the River Ravi right where Jammu, Punjab, and Chamba territories intersect. Its painters depicted religious scenes with circular enclosures of forest and flowering trees, often surrounding a pond with floating lotus leaves and flowers, opening up to a river in the foreground. Further upstream, Chamba state drains the Ravi river headwaters. Its paintings feature more elaborate forest scenes in waterfront locations. Some depict Rama, Sita, and Lakshman in their forest hermitage. Many celebrate Radha and Krishna's love in beautiful forest settings along rivers, as well as rajas and ranis emulating those passions in formal garden pavilions like those of the Mughals.

To the southeast, the Beas River flows through the former Mandi, Kangra, and Guler states. Mandi and Kangra paintings feature images of Radha longing for Krishna in forests and waterscapes. Downstream on the Beas, Guler artists produced paintings that have abundant images of lakes, rivers, and ponds. Many feature women swimming, including the Punjabi folktale of Sohni floating dangerously across the river to meet her lover Mahiwal. One Guler painting depicts an irrigated garden scene with a Persian wheel and gardeners tending flower parterres in a large enclosure (Archer 1973, 111, 112). For the most part, Punjab hill paintings emphasized forest landscapes with free-flowing rivers and ornamental waterworks, which complement the intensively irrigated plains below.

In both areas—hills and plains—pre-nineteenth-century irrigation practices relied on streams, wells, and inundation canals. Water-lifting devices included buckets, Persian wheels, chains of pots, and inclined ramps drawn by animals (Crooke 1989). The early social histories of these irrigated gardens across the Punjab remain to be written. They are crucial for constructing a cultural approach to water management from the ground up that would draw upon all of the humanities, along with the historians who have contributed so much to date (e.g., Gilmartin 2015).

The culture of colonial canal irrigation, by comparison, involved extensive manipulation of social and environmental relationships, coupled with irrigation projects on a monumental scale. The first of these projects began with strategic cartographic surveys of the river and region by explorers such as Alexander Burnes (1834), who praised the lush irrigated gardens along the river in Multan. These river surveys contributed to the aggressive annexation of Punjab and Sindh, after which the British took a combination of military and civilian approaches to irrigated gardens by constructing Soldier Baghs, Company Baghs, botanical gardens, and canal plantations of great length and diversity. For example, the early Bari Doab Canal reports mentioned that, notwithstanding a decade of political disturbance, cultivation of the region was almost garden-like in character (Crofton, Dyas, and Napier 1850). Canal planner Richard Baird Smith (1849) wrote about the agricultural potential of the Punjab, noting its carefully irrigated and well-tended garden-like character. It is interesting to note that when California engineers sought models for developing the Central Valley, discussed in Faletti's chapter in this volume, they viewed colonial Punjab as a promising precedent (Wescoat 2000).

The vast canal irrigation system constructed in Punjab during the late nineteenth and twentieth centuries settled decommissioned soldiers and pastoral peoples on newly irrigated lands to increase political control and revenue (Gilmartin 2015). The British manipulated tribal customary law and power relations when laying out settlements and allocating land grants. Unlike gardens in the upper basin, these early canal irrigation projects sought to spread water as far and as thinly as possible to maximize social and territorial control.

To accomplish this aim, irrigators received relatively low water entitlements in rotations or "turns" known as *waribandi* (Wescoat 2013). Previous research has examined the (in)efficiency and (un)fairness of these water deliveries in relation to entitlements, frequently noting the low water deliveries to low status farmers at the tail ends of distributaries. It is speculative but inspiring to consider the cultural bases for regarding a turn (*waar*) not so much as a right as a form of duty or even sacrifice to others, as in the Punjabi Sufi poem titled "Sammi waar." Sufi literature has many tales of sacrifice (ایثار) in which a small amount of water is shared among dying companions stranded in the desert or on a battlefield. The companions keep sacrificing (واری) their turn (وار) for others, and the cup of water is thus never entirely consumed. In an ideal world, water users of all types and at all levels might be imagined to واری their وار for others. Although the social science literature records many examples of the opposite behavior, to our knowledge few have searched for cultural evidence of such sacrifice.

Concurrent with the expansion of canal irrigation in the mid-nineteenth century was the establishment of the Agri-Horticultural Society of the Punjab. As its title indicates, it aimed to create a bridge between field and garden cultivation (Kerr 1976; Rehman 2014). Punjab was not the only branch, but it was an active one, originating just two years after annexation of the province by order of Governor-General Dalhousie, who lamented the lack of forest, fruit, and shrub cover in Punjab's agrarian landscape (Kerr 1976). Dalhousie gave directions to line the banks of the Bari Doab Canal with tree plantings and encouraged village communities to set aside lands for this purpose. While largely functional in purpose, the cultural motivations for corridors of road and canal bank tree plantings soon after annexation reflected an imperial vision of landscape transformation and bounty.

Following partition and independence in 1947, postconflict reconstruction included cooperative "garden colonies" in irrigated districts of East Punjab to promote fruit production, especially citrus. The duty of water (water allocation per acre) was increased for these valuable garden crops (Randhawa 1954). While the culture of irrigated gardens in Punjab has a rich and diverse history, its breadth, depth, and salience for the basin's past and future have yet to be fully embraced.

5. Irrigated Gardens of the Lower Indus Corridor and Delta

The lower Indus basin is an arid landscape with broad floodplains and extensive irrigation. It may seem the least garden-like region in character, but only if one fails to search and see (Haines 2013; Pithawala 1937). Some colonial writers wrote despondently about the region as "the unhappy valley," and indeed the lower Indus faces some of the greatest livelihood challenges of waterlogging, salinity, drainage, flooding, oppression, and deltaic deterioration.

But the lower Indus River channel and plains have also been envisioned at various places and times as irrigated gardens. For example, a leading scholar in Islamic studies, Annemarie Schimmel (1999), cites modern poets and Sufi saints on the beauty and bounty of the lower Indus, including its barrages: "What a plenitude of fruits and vegetables can now be grown in the country! From carrots, garlic, corn, watermelon and sugarcane to wholesome plants that help cows to produce more milk—there is 'relief after grief.' . . . Not only dozens of hitherto unknown

FIGURE 8.3. Orangi Pilot Project self-help wastewater sewerage and community development. Photo courtesy James Wescoat.

vegetables and fruits can now be harvested—there is also a variety of birds which begin to nest in the fresh greenery, among them crows and doves, Meena birds and ring doves and many others" (413).

Upstream near Sukkur barrage, Larkana district has long been known as the "garden of Sindh" and the "Eden of Sindh" for its bountiful rice cultivation (Holmes 1968). The *Globe Encyclopedia* described Larkana in 1878 as "situated in the most fertile part of the province, and from its gardens and tree-lined roads has been called the 'Eden of Scinde'" (Ross 1878, 56). During the colonial period, Sindh was famous for its fruit and vegetable cultivation, as well as for its rice, wheat, and cotton. Aitken (1907, 36–38) listed dozens of garden fruits and their performance in the province.

Just above the delta, the Mughal capital at Thatta was famous for its flower and fruit gardens (especially pomegranates) (Aitken 1907, 238). The successor capital at Karachi also had a reputation for fine gardens at the turn of the last century: "along the Lyari River and as far as Malir, there are gardens, owned by Khojas or Memons, which supply the market with all the standard fruits" (238; though see the critical contemporary perspective of Ginn 2018). Despite all of its stresses, the vast low-income Orangi neighborhood in Karachi has kitchen gardens, tree plantings, and a plant nursery as part of its wastewater infrastructure and community development program ("Orangi Pilot Project: NGO Profile" 1995) (figure 8.3).

Even the deteriorating Indus delta has had its gardens. Medieval geographers like al-Idrisi mentioned towns surrounded by gardens, one of them known as Jun on a branch of the lower Indus (Haig 1894). "Clay vessels, leather bags, and other receptacles were filled from the surface water bodies and were carried by man to the required places. The amount of water thus raised was sufficient for the small farms and gardens common to early farmers" (Rahman 1960, 78).

SYNTHESIS AND IMPLICATIONS FOR INDUS RIVER BASIN MANAGEMENT

We have followed the Indus from its vast headwaters in the Kabul, Jhelum, Chenab, and Beas-Sutlej watersheds through the plains and down to the delta. We are now in a position to draw these threads together in a perspective on irrigated gardens of the Indus, and imagine the entire Indus as a landscape of irrigated gardens. Indeed, five humanistic threads seem to tie these irrigated garden cultures together:

Deep historical continuity. The Indus basin has deep garden history and heritage, originating in the earliest centers of civilization at Harappa and continuing up to new internet forums in Pakistan that feature hundreds of garden webpages.

Basin-wide geographical extent. We have shown that inspired and inspiring irrigated gardens extend from the upper reaches of the Indus main stem and all of its tributaries, through the broad plains of Punjab and Sindh, into the estuaries of the Indus River delta. Many of these gardens are fed by small tributary streams, and some rely on wells, while still others draw upon major river channels and floodplains that connect the Indus as a basin.

Diverse types of irrigated gardens. The diversity of irrigated gardens ranges from rice fields to orchards and flower gardens. These extend from the smallest kitchen gardens to the entire irrigated basin. The denotation of the irrigated garden thus encompasses remote rural villages, megacities, and expanding suburbs.

Shared meanings and values. Notwithstanding this diversity, our interpretation reveals several widely held connotations of irrigated gardens. There are important variations, to be sure, but well-watered gardens are associated with the creation of the world, its sacredness, human stewardship, and rewards for those who have done good. The best gardens reflect the virtues and dignity of lives well lived.

Visions for a common future. Collective understanding, be it in the form of myths, stories, and legends or as shared values reflected in gardens, provides ideals for envisioning a common future for the basin. Although not explored in this essay, futuristic imagination that features societal interactions with new technologies (Muhammad 2016) and new environmental challenges (e.g., climate change) may be explored using the garden as a metaphor. The good irrigated gardens of this world are indeed signs of beneficence and of what is possible.

Collectively, these five aspects of irrigated gardens in the Indus offer culturally inspired and inspiring models of conscientious waterscapes that embrace and go beyond the accomplishments of twentieth-century water management. No river basin model currently aspires to these humanistic ideals or offers real and imagined examples of their potential. Mainstream models of river basin management speak to other social objectives. Colonial approaches combined the aims of economic efficiency and sociopolitical control (Gilmartin 2015). Postcolonial accounts remain focused on the politics and geopolitics of water development (Amrith 2018). Modern multi-objective river basin planning has addressed trade-offs among economic, ecological, and equity goals in ways that promoted economic growth and development. Overall, however, these modern approaches did not aim high enough in cultural terms. They sought to enhance material production to achieve social goals and mitigate the harms of water development projects. While those aims represent an advance over the current situation of inefficiency, inequity, and environmental deterioration, they are much less than what is possible with a cultural approach on a river basin scale.

Critical assessments have underscored the deficiencies of modern water management (Mustafa 2013). It is not just inefficiency: unfair competition, unbridled capital accumulation, and oppressive power relations are widespread. Conflict is one of the most prevalent themes in water resources research in the Indus and elsewhere (Adeel and Wirsing 2017; Amrith 2018; Haines 2016). The history of water resource management is replete with examples of social domination, oppression, and inequity (Naqvi 2013). The humanities have also shed light on these dark themes in garden history (Wescoat 1991; 1995).

In place of the illusory goal of optimizing economic development or romanticizing traditional landscapes, the cultural approach espoused here helps articulate the "purpose" of a river basin's existence, the "meaning" of its people's aspirations, the moral dangers of mismanagement, and the aesthetic prospects of the irrigated garden model. As demonstrated here, a cultural approach can identify places on the ground, throughout history, and throughout the basin, that manifest the beauty and meaning of irrigated gardens in theory and in practice. These irrigated gardens constitute a mosaic and network of inspired and inspiring water flows in the Indus River basin. Collectively, they point toward larger processes of irrigated garden development at the scale of canal commands, provinces, nation-states, the international river basin, and beyond. They remind one of the paradise gardens at the end of time, and how they guide those who strive to have faith and do good work in the Indus basin today.

ACKNOWLEDGEMENTS

This paper has benefited from many readings and responses, especially from the editors and their colleagues at the University of California-Merced where the Water and Humanities group challenged us to look beyond the political boundaries of Pakistan. The result is a stronger and broader aspiration for a basinwide perspective. That perspective was further fostered by a conference at the International Institute of Applied Systems Analysis (IIASA) in Vienna in 2018 where scientists from all of the riparian states exchanged ideas in the best spirit of international collaboration and gave encouragement for pursuing a garden approach. We are grateful to our colleague Dr. Afreen Siddiqi for her ideas and feedback. This paper had its origins in a project on Mughal gardens generously funded by the Smithsonian Institution in the 1980s, in collaboration with colleagues at the University of Engineering and Technology-Lahore, which led to many subsequent explorations of the meaning and prospects of water gardens in South Asia. Among those, we thank Shahrukh Hamid and Kunwar Schahzeb of LUMS Media Lab for their creative work on an Indus *River Garden* film.

REFERENCES

Acciavati, Anthony. 2015. *Ganges Water Machine: Designing New India's Ancient River*. Applied Research + Design.

Adeel, Zafar, and Robert G. Wirsing, eds. 2017. *Imagining Industan: Overcoming Water Insecurity in the Indus Basin*. Dordrecht: Springer.

Agarwal, Arun, and K. Sivaramakrishnan, eds. 2000. *Agrarian Environments: Resources, Representations, and Rule in India*. Durham, NC: Duke University Press.

Ahmad, Masood, and Kutcher, Gary P. 1992. *Irrigation Planning with Environmental Considerations: A Case Study of Pakistan's Indus Basin*. World Bank Technical Paper no. 166. Washington, DC: The World Bank.

Aitchison, James E.T. 1869. *A Catalogue of the Plants of the Punjab and Sindh*. London: Taylor & Francis.

Aitken, Edward Hamilton. 1907. *Gazetteer of the Province of Sind*. Karachi: "Mercantile" Steam Press.

Amrith, Sunil. 2018. *Unruly Waters: How Rains, Rivers, Coasts, and Seas Have Shaped Asia's History*. New York: Basic Books.

Angchok, Dorjey, and Premlata Singh. 2006. "Traditional Irrigation and Water Distribution System in Ladakh." *Indian Journal of Traditional Knowledge*. 5 (3): 397–402.

Apte, Vaman Shivram. 1957–59. *The Practical Sanskrit-English Dictionary*. Revised and enlarged ed. Poona: Prasad Prakashan.

Archer, William G. 1973. *Indian Paintings from the Punjab Hills: A Survey and History of Pahari Miniature Painting*. London: Sotheby Parke Bernet.

Bianca, Stefano, ed. 2005. *Karakorum: Hidden Treasures in the Northern Areas of Pakistan*. Aga Khan Trust for Culture.

Briscoe, John, and Usman Qamar. 2005. *Pakistan's Water Economy: Running Dry*. Washington, DC: World Bank.

Buache, Philippe. 1757. "Carte physique de la mer des Indes." www.wdl.org/en/item /18893/.

Burnes, Alexander. 1834. *Travels into Bokhara: Being the Account of a Journey from India to Cabool, Tartary, and Persia; Also, Narrative of a Voyage on the Indus [. . .]*. London: John Murray.

Casimir, Michael J. 2021. *Floating Economies: The Cultural Ecology of the Dal Lake in Kashmir, India*. New York: Berghahn.

Cheema, M.J.M., W.G.M. Bastiaanssen, and M.M. Ruttena. 2011. "Validation of Surface Soil Moisture from AMSR-E Using Auxiliary Spatial Data in the Transboundary Indus Basin." *Journal of Hydrology* 405 (1–2): 137–49.

Chellany, Brahma. 2013. *Water: Asia's New Battleground.* Washington, DC: Georgetown University Press.

CRCI (Cultural Resource Conservation Initiative). 2009. Comprehensive Conservation Management Plan of the Red Fort, Delhi. Vol. 1. New Delhi: Archaeological Survey of India and CRCI. https://asi.nic.in/comprehensive-conservation-management-plan-for -red-fort-delhi-ccmp/.

Crofton, J., J.H. Dyas, and Robert Cornelius Napier. 1850. *Reports on the Baree Doab Canal Project.* London: printed by Day and Son.

Crooke, William. 1989. *A Glossary of North Indian Peasant Life.* New Delhi: Oxford University Press.

Drew, Frederic. 1874. "The Upper Indus Basin." *Geological Magazine* 1 (2): 94.

Duloy, John H., and Gerald T. O'Mara. 1984. *Issues of Efficiency and Interdependence in Water Resource Investments: Lessons from the Indus Basin of Pakistan.* World Bank Staff Working Paper no. 665, Washington, DC: World Bank.

Falk, Harry. 2009. "The Pious Donation of Wells in Gandhara." In *Prajñādhara—Essays on Asian Art, History, Epigraphy and Culture in Honour of Gouriswar Bhattacharya*, edited by Gerd J.R. Mevissen and Arundhati Banerji, 23–36. New Delhi: Kaveri Books.

Gilmartin, David. 2015. *Blood and Water: The Indus River Basin in Modern History.* Berkeley: University of California Press.

Ginn, Franklin. 2018. "Plant Politics in Karachi." *Arcadia* no. 12 (Spring). DOI: https://doi .org/10.5282/rcc/8318.

Gole, Susan. 1989. *Indian Maps and Plans.* Delhi: Manohar.

Greenwood, George. 1874. "The Upper Indus Basin." *Geological Magazine* 1 (1): 45.

Gulhati, Niranjan Das. 1973. *Indus Waters Treaty: An Exercise in International Mediation.* Bombay: Allied Publishers.

Haig, Malcolm Robert. 1894. *The Indus Delta Country: A Memoir, Chiefly on Its Ancient Geography and History.* London: K. Paul.

Haines, Daniel. 2013. *Building the Empire, Building the Nation: Development, Legitimacy, and Hydro-Politics in Sind, 1919–1969.* Karachi: Oxford University Press.

———. 2016. *Rivers Divided: Indus Basin Waters in the Making of India and Pakistan.* Karachi: Oxford University Press.

Holmes, D.A. 1968. "The Recent History of the Indus." *The Geographical Journal* 134 (3): 367–82.

Hussain, Ijaz. 2017. *Indus Waters Treaty.* Karachi: Oxford University Press.

Hussain, Shafqat. 2015. *Remoteness and Modernity: Transformation and Continuity in Northern Pakistan.* New Haven: Yale University Press.

Ibbetson, Denzil, and Edward Maclagan. 1911–19. *A Glossary of the Tribes and Castes of the Punjab and Northwest Frontier Province.* 3 vols. Lahore: Aziz.

ICIMOD. 2018. *Upper Indus Basin-Network (UIB-N) Workshop for Enhancing Science-Based Regional Cooperation, April 24–25, 2018.* Kathmandu, Nepal: ICIMOD Proceedings 2018/5.

Inden, Ronald. 2008. 'Kashmir as Paradise on Earth." In *The Valley of Kashmir: The Making and Unmaking of a Composite Culture?*, edited by Aparna Rao, 523–62. New Delhi: Oxford University Press.

INTACH (Indian National Trust for Art and Cultural Heritage, Jammu and Kashmir branch). 2014. *Shehr e Kashmir*. 2 vols. Srinagar: INTACH.

IWMI (International Water Management Institute). N.d. "A Brief History of IWMI." www .iwmi.cgiar.org/about/who-we-are/iwmi-history/. Accessed November 11, 2018.

Jahangir. 1999. *The Jahangirnama: Memoirs of Jahangir, Emperor of India*. Translated by Wheeler M. Thackston. Washington, DC: Freer Gallery of Art and Arthur M. Sackler Gallery, Smithosonian Institute; New York: Oxford University Press.

Jamison, Stephanie W., and Joel P. Brereton, trans./eds. 2017. *The Rigveda*. 3 vols. Oxford: Oxford University Press.

Johnson, Sam H., Alan C. Early, and Max K. Lowdermilk. 1977. "Water Problems in the Indus Food Machine." *Water Resources Bulletin* 13 (6): 1253–68.

Kango, A.M.H. 1997. "Breakdown of Indus Food Machine." http://wgbis.ces.iisc.ernet.in /envis/doc97html/envsdpak514.html. Accessed May 7, 2021.

Kaul, Shonaleeka. 2018. *The Making of Early Kashmir: Landscape and Identity in the Rajatarangini*. New Delhi: Oxford University Press.

Kerr, Ian. 1976. "The Agri-Horticultural Society of the Punjab, 1851–71." In *Punjab Past and Present: Essays in Honour of Dr. Ganda Singh*, edited by Harbans Singh and N.G. Barrier, 252–72. Patiala: Punjabi University.

Khan, Asif, Keith S. Richards, Geoffrey T. Parker, Allan McRobie, and Biswajit Mukho-padhyay. 2014. "How Large Is the Upper Indus Basin? The Pitfalls of Auto-Delineation Using DEMs." *Journal of Hydrology* 509: 442–53.

Koch, Ebba. 2008. "My Garden Is Hindustan: The Mughal Emperor's Realization of a Political Metaphor." In *Middle East Garden Traditions: Unity and Diversity*, edited by Michel Conan, 460–75. Washington, DC: Dumbarton Oaks.

Kreutzmann, Hermann, ed. 2000. *Sharing Water: Irrigation and Water Management in the Hindukush, Karakorum, Himalaya*. Karachi: Oxford University Press.

———, ed. 2006. *Karakorum in Transition*. Karachi: Oxford University Press.

Leighly, John, ed. 1963. *Land and Life: A Selection of Writings by Carl Ortwin Sauer*. Berkeley: University of California Press.

Lenton, Roberto, and Mike Muller. 2009. *Integrated Water Resources Management in Practice: Better Water Management for Development*. New York: Routledge.

Lilienthal, David. 1951. "Another 'Korea' in the Making?" *Collier's Weekly*, August 4, 1951, 22–23, 56–58.

Losty, Jeremiah P. 2017. *A Mystical Realm of Love: Pahari Paintings from the Eva and Konrad Seitz Collection*. London: Ad Ilissum.

Maass, Arthur, et al., eds. 1962. *Design of Water Resource Systems*. Cambridge, MA: Harvard University Press.

MacDonnell, Arthur Anthony. 1929. *A Practical Sanskrit Dictionary with Transliteration, Accentuation, and Etymological Analysis Throughout*. London: Oxford University Press.

Malhotra, Anshu, and Farina Mir, eds. 2012. *Punjab Reconsidered: History, Culture, Practice*. New Delhi: Oxford University Press.

McCrindle, John Watson. 1896. *The Invasion of India by Alexander the Great, as Described by Arrian, Q. Curtius, Diodoros, Plutarch and Justin*. Westminster: A. Constable.

Meadows, Azra, and Peter Meadows, eds. 1999. *The Indus River: Biodiversity, Resources, Humankind*. Linnean Society symposium. Karachi: Oxford University Press.

Mehta, Lyla, Bill Derman, and Emmanuel Manzungu, eds. 2016. Special issue, "Flows and Practices: The Politics of Integrated Water Resources Management (IWRM) in Southern Africa." *Water Alternatives* 9 (3).

Merrey, Douglas J. 1986. "The Local Impact of Centralized Irrigation Control in Pakistan: A Sociocentric Perspective." In *Irrigation Management in Pakistan: Four Papers*, edited by Douglas J. Merrey and James M. Wolf. Digana Village, Kandy, Sri Lanka: International Irrigation Management Institute.

Michel, Aloys. 1967. *The Indus Rivers: A Study of the Effects of Partition*. New Haven: Yale University Press.

Mitchell, Don. 1996. *The Lie of the Land*. Minneapolis: University of Minnesota Press.

Mitchell, W.J.T. 2002. *Landscape and Power*. 2nd ed. Chicago: University of Chicago Press.

Mollinga, Peter, Kusum Athukorala, and Ajaya Dixit, eds. 2006. *Integrated Water Resources Management: Global Theory, Emerging Practice and Local Needs*. New Delhi: Sage Publications.

Mughal, Mohammed Rafique. 1997. *Ancient Cholistan: Archaeology and Architecture*. Lahore: Ferozsons.

Muhammad, Abubakr. 2016. "Managing River Basins with Thinking Machines." In *2016 IEEE Conference on Norbert Wiener in the 21st Century (21CW)*, 1–6. DOI: https://doi.org/10.1109/NORBERT.2016.7547453.

Mustafa, Daanish. 2013. *Water Resource Management in a Vulnerable World: The Hydrohazardscape of Climate Change*. London: I.B. Tauris.

Naqvi, Saiyid Ali. 2013. *Indus Waters and Social Change: The Evolution and Transition of Agrarian Society in Pakistan*. Karachi: Oxford University Press.

"Orangi Pilot Project: NGO Profile." 1995. *Environment and Urbanization* 7 (2): 227–36.

Ostrom, Elinor. 1990. *Governing the Commons: The Evolution of Institutions for Collective Action*. Cambridge: Cambridge University Press

———. 1993. "Design Principles in Long-Enduring Irrigation Institutions." *Water Resources Research* 29 (7): 1907–12.

Pali Text Society. 1921–25. *The Pali Text Society's Pali-English Dictionary*. Chipstead.

Pandit, Ranjit Sitaram. 1935. *Rajatarangini: The Saga of the Kings of Kashmir*. New Delhi: Sahitya Akademi.

Parihar, Subhash. 1989. *Some Aspects of Indo-Islamic Architecture*. New Delhi: Abhinav.

Parodi, Laura. 2021 (forthcoming). "Kabul, a Forgotten Mughal Capital: Gardens, City, and Court at the Turn of the Sixteenth Century." *Muqarnas* 38.

Parveen, Sitara, Matthias Winiger, Susanne Schmidt, and Marcus Nüsser. 2015. "Irrigation in Upper Hunza: Evolution of Socio-Hydrological Interactions in the Karakoram, Northern Pakistan." *Erdekunde* 69 (1): 69–85.

Petruccioli, Attilio, ed. 1985. "Water and Architecture." Special issue of *Environmental Design: Journal of the Islamic Environmental Design Research Centre* 2.

Phillimore, R.H. 1952. "Three Indian Maps." *Imago Mundi* 9 (1): 111–13.

Pithawala, M. 1937. *A Geographical Analysis of the Lower Indus Basin (Sind)*. Bombay: M.B. Pithawala.

Platts, John Thompson. 1884. *A Dictionary of Urdu, Classical Hindi, and English*. London: W.H. Allen.

Rahman, Mushtaq-ur. 1960. "Irrigation and Field Patterns in the Indus Delta." PhD diss., Louisiana State University.

Rajamani, Imke, Margrit Pernau, and Katherine Schofield, eds. 2018. *Monsoon Feelings: A History of the Emotions in Rain*. New Delhi: Niyogi Books.

Rajapurohit, B.B. 2012. *Grammar of Shina Language and Vocabulary (Based on the Dialect Spokenaround Dras)*. www.yumpu.com/en/document/view/49143939/grammar-of-shina-language-and-vocabulary.

Randhawa, Mohindar S. 1954. "Out of the Ashes: An Account of the Rehabilitation of Refugees from West Pakistan in Rural Areas of East Punjab." Public Relations Dept., Punjab. https://archive.org/details/dli.csl.8772.

Rao, Aparna, ed. 2008. *The Valley of Kashmir: The Making and Unmaking of a Composite Culture?* Delhi: Manohar.

Ray, Sugata, and Venugopal Maddipati, eds. 2020. *Water Histories of South Asia: The Materiality of Liquescence*. New York: Routledge.

Rehman, Abdul. 1996. "The Gardens of Peshawar." In *The Mughal Garden: Interpretation, Conservation and Implications*, edited by Mahmood Hussain, Abdul Rehman, and James L. Wescoat Jr., 89–92. Lahore: Ferozsons.

Rehman, Nida. 2014. "Description, Display and Distribution: Cultivating a Garden Identity in Late Nineteenth-Century Lahore." *Studies in the History of Gardens & Designed Landscapes* 34 (2): 176–86.

Revelle, Roger. 1985. Interview with Dr. Roger Revelle. May 15–16, 1985, University of California, San Diego, 25th Anniversary Oral History Project. https://library.ucsd.edu/speccoll/siooralhistories/2010-44-Revelle.pdf.

Ross, John, ed. 1876. *The Globe Encyclopedia*. Vol. 4. Edinburgh: Thomas C. Jack, Grange Publishing Works.

Schimmel, Annemarie. 1999. "The Indus—River of Poetry." In *The Indus River: Biodiversity, Resources, Humankind*, edited by Azra Meadows and Peter S. Meadows, 409–15. Karachi: Oxford University Press.

Schön, Donald. 1979. *Generative Metaphor: A Perspective on Problem-Setting in Social Policy*. Cambridge: Cambridge University Press.

Sharma, Sunil. 2017. *Mughal Arcadia: Persian Literature in an Indian Court*. Cambridge, MA: Harvard University Press.

Singh, Lakhwinder, Kesar Singh Bhangoo, and Rakesh Sharma. 2018. *Agrarian Distress and Farmer Suicides in North India*. New Delhi: Routledge.

Smith, Richard Baird. 1849. *Agricultural Resources of the Punjab: Being a Memorandum on the Application of the Waste Waters of the Punjab to Purposes of Irrigation*. London: Smith, Elder.

Sri Granth. N.d. "Garden." www.srigranth.org/. Accessed June 3, 2019.

Stein, Marc Aurel. 1900. *Rajatarangini*. Vol. 2, *Geographical Memoir*. Westminster: Archibald Constable.

Swynnerton, Charles. 1987. *Folk Tales from the Upper Indus*. Islamabad: Lok Virsa.

Wahi, Tripta. 2014. "Irrigation and Labor in Precolonial Punjab." *Proceedings of the Indian History Congress* 75, Platinum Jubilee: 400–408.

Wallach, Bret. 1996. *Losing Asia: Modernization and the Culture of Development*. Baltimore: Johns Hopkins University Press.

Weber, Steven A. 1999. "Seeds of Urbanism: Paleoethnobotany and the Indus Civilization." *Antiquity* 73, 813–26.

Weber, Steven A., and William Raymond Belcher, eds. 2003. *Indus Ethnobiology*. Lanham, MD: Lexington Books.

Wescoat, James L., Jr. 1989. "Picturing an Early Mughal Garden." *Asian Art* 2: 59–79.

———. 1991. "Gardens of Conquest and Transformation: Lessons from the Earliest Mughal Gardens in India." *Landscape Journal* 10 (2): 105–14.

———. 1995. "From the Gardens of the Qur'an to the Gardens of Lahore." *Landscape Research* 20: 19–29.

———. 2000. "Wittfogel East and West: Changing Perspectives on Water Development in South Asia and the US, 1670–2000." In *Cultural Encounters with the Environment: Enduring and Evolving Geographic Themes.*, edited by Alexander B. Murphy and Douglas L. Johnson, 109–32. Lanham, MD: Rowman & Littlefield.

———. 2012. "The Indus River Basin as Garden." *Die Gartenkunst*, 24 (1): 19–33.

———. 2013. "Reconstructing the Duty of Water: A Study of Emergent Norms in Socio-Hydrology." *Hydrology and Earth System Sciences* 17 (12): 4759–68.

———. 2017. "Water Resources and Hydrological Management." *International Encyclopedia of Geography*. https://doi.org/10.1002/9781118786352.wbieg0620.

———. 2021. "Conservation of Indo-Islamicate Water Experience." In *Heritage Conservation in Postcolonial India, Approaches and Challenges*, edited by Manish Chalana and Ashima Krishna, 235–51. New Delhi, Routledge.

Wescoat, James L., Jr., Sarah Halvorson, and Daanish Mustafa. 2000. "Water Management in the Indus Basin of Pakistan: A Half-Century Perspective." *International Journal of Water Resources Development* 16: 391–406.

Wescoat, James L., Jr., Abubakr Muhammad, and Afreen Siddiqi. 2018. "Socio-Hydrology of Channel Flows in Complex River Basins: Rivers, Canals, and Distributaries in Punjab, Pakistan." *Water Resources Research* 54: 464–79.

Wiener, Norbert. 1961. *Cybernetics or Control and Communication in the Animal and the Machine*. Rev. ed. Cambridge, MA: MIT Press.

Wood, Geoffrey, Abdul Malik, and Sumaira Sagheer, eds. 2006. *Valleys in Transition: Twenty Years of AKRSP's Experience in Northern Pakistan*. Karachi: Oxford University Press.

World Bank. 2010. *Scoping Strategic Options for Development of the Kabul River Basin: A Multisectoral Decision Support System Approach*. Washington, DC.

World Bank. 2017. *Pakistan—Punjab Irrigated Agriculture Productivity Improvement Program Project*. Washington, DC.

Wright, Rita P. 2010. *The Ancient Indus: Urbanism, Economy and Society*. Cambridge: Cambridge University Press.

Yazdi, Sharaf ad-Din Ali. 1976 (reprint). "Zafarnama." In *The History of India as Told by Its Own Historians*, edited by H.M. Elliott and John Dowson. Lahore: Islamic Book Service.

Yu, Winston, et al. 2013. *The Indus Basin of Pakistan: The Impacts of Climate Risks on Water and Agriculture*. Washington, DC: World Bank.

Zutshi, Chitralekha. 2014. *Kashmir's Contested Pasts: Narratives, Sacred Geographies and the Historical Imagination*. New Delhi: Oxford University Press.

Leadership in Principle

Uniting Nations to Recognize the Cultural Value of Water

Veronica Strang

In 2016, the United Nations confronted alarming projections about impending water shortages worldwide, suggesting a potential 40 percent shortfall in water availability by 2030 (United Nations High Level Panel on Water 2018, 7). Concerns about water scarcity had been growing over the previous decade: the United Nations Development Programme suggested in its 2006 report that water use was rising at twice the rate of population growth and that a quarter of the world's population faced impending problems with water scarcity (UNDP 2006). By 2016, the scale of the problem had become painfully evident to many state governments and national and international NGOs. It was clearly a matter of increasing urgency to provide global leadership focused specifically on water. The UN therefore appointed a High Level Panel on Water to address these issues and to develop some new principles for water to underpin its recently established sustainability goals. Such panels are led by what the UN calls *sherpas*, underscoring the expectation that they will provide a lofty overview. The panel's remit was to encourage heads of state to develop more sustainable policies and practices in relation to water governance, management, and use, in particular by rethinking how water is valued. Moving beyond a largely economic or technical conversation required the panel to make greater use of the social sciences' and humanities' capacities to address the complex social and cultural issues that attend diverse evaluations of water.

This chapter examines that process and considers how such "top-down" endeavors might engender changes in policy and open up new ways of engaging with water, both among national leaders and at a grassroots level. It therefore engages with research exploring the relationships between local communities, major governmental and nongovernmental organizations, and international networks.

It is now some time since anthropology began to "study up," as social anthropologist Laura Nader recommended, to consider the powerful organizations and wider networks with which local communities engage in negotiating water issues (1969). Much good ethnography has been written about organizations (Corsín Jiménez 2007; Garsten and Nyqvist 2015; Nash 1979) and on state bureaucracies (Herzfeld 1992). Fajardo's analysis of Malalos, Philippines, in this volume exemplifies the utility of this direction. There is also a range of work from within anthropology and development studies considering the nearly forty thousand NGOs to have emerged in recent decades. With varying degrees of criticism, analyses have considered the role of such organizations, and in particular international NGOs: as aid and development agencies, as mechanisms for the dissemination of neoliberal values, and as attempts to access markets and resources (Arce and Long 1999; Crewe and Axelby 2012; Escobar 2005; Lewis and Mosse 2006). Such work has highlighted the reality that all such organizations contain important social and professional networks that, via complex linkages, intersect dynamically at a variety of scales (Latour 2005; Scott 2000). In this volume, for example, Wescoat and Muhammad examine the importance of allowing cultural values to lead water policy, management, and systems conversations in the Indus River Basin.

In effect, in relation to water, as well as other global policy issues, NGO networks have generated national and international elites, empowered by political influence, access to resources, and technical or scientific expertise (Allen 2018; Lashaw, Vannier, and Sampson 2017). As private organizations, NGOs have highly varied aims and motivations. Their activities may be complementary to state efforts, but they may also displace state agencies (Leve and Karim 2001), or challenge state orthodoxies and act as catalysts for change and transformation (Korten 1990). They can enable the wider embedding of neoliberal ideologies or support challenges to dominant norms and countermovements. Thus, Tsing observes that, in some contexts, they have the potential to provide alternatives to state authority and promote environmental justice (2005). The Earth Law Centre and the Global Alliance, for example, are lobbying for environmental legislation aimed at improving nonhuman rights, including the rights of rivers and ecosystems. Other NGOs and networks are more concerned with conventional development goals. For instance, as an industry-based organization, Water UK aims to bring expertise from British water companies to support the provision of clean water and sanitation in countries where such current technologies for doing so are insufficient. But at the same time, in the decade following water privatization in Britain in 1989, some of its member companies were simultaneously (and very profitably) involved in offering expertise in water privatization to the governments of these nations. Many such privatizations have been extremely problematic, leading to major conflicts, and in the last decade there has been a countertrend towards reversions to state provision (Hoedeman et al. 2013; Kishi-moto, Lobina, and Petitjean 2015).

In this complex global mix of governments and international organizations, the United Nations has a vital responsibility to try to bring diverse actors together. Established in 1945 as an intergovernmental organization, its first and foremost role, following the devastation of World War II, was to maintain peace and order between nations (Ghali 1992). Today the UN contains 193 member states, and its mission has expanded, logically, into areas that are vital to peace and order, such as international justice, human rights, health, and the conservation of cultural heri-tage. Its interests in health include long-standing concerns about the provision of clean water and sanitation, and there is strong recognition that health, social order, and stability also depend on water (and thus food) security. The UN's global efforts to address water issues have been assisted by a variety of UNESCO programs, such as the International Ecohydrology Programme, its extensive work on water and cultural heritage, and the establishment of the UNESCO-IHE Delft Institute. The UN's agenda for 2030, issued in September 2015, defined seventeen Sustainable Development Goals, and its more recent efforts to compose some new Principles for Water acknowledged the centrality of water in all of these.

In relation to water, as with other global policy issues, the UN has a range of formal and informal mechanisms for articulating its goals and bringing state and nonstate actors together. Its activities are iterative: consultation generates some agreement on ways forward, preliminary plans are set out in reports, further con-sultation follows, and the process is ongoing. At various junctures, agreement is reached between its members, and declarations, goals, and principles are formally established. Inevitably, the contributors to the conversations facilitated by the UN have varying degrees of power and influence, and there is a tendency for dominant ideologies, understandings, and discourses to prevail. Just as more powerful politi-cal actors tend to have the loudest voice in debates, so too do the more influential academic disciplines: thus, UN conversations about water have long been led by STEM disciplines and economics. However, this began to change in the 2000s, with more input from the social sciences into the International Ecohydrology Pro-gramme, into UNESCO's work on water and cultural diversity (Johnston et al. 2012), and in the founding of the International Water History Association in 2001, which Faletti mentions in this volume. This involvement helped, in particular, to highlight the variations in people's relationships with water, and in the geopolitics of different national and cultural contexts.

Recent advances in digital technology have also widened the pool of potential participants in UN activities. For the first time in human history, it has become possible to conduct global conversations in real time, Skyping or Zooming in peo-ple who might otherwise find it difficult to make it to formal meetings. The virtual networks enabled by social media mean that many previously marginalized groups have greater capacity to speak for themselves and to each other (Miller et al. 2016). Some, such as Indigenous communities, have created their own international networks that have significantly increased their abilities to compare notes, to articulate their shared beliefs and values, and to be heard in global fora. Thus the panel's meetings in The Hague to compose the Principles for Water included—in person and via digital media—not only governmental, NGO, and economic advisers, but also a wider range of disciplinary experts and representatives from diverse cultural contexts.

The potential to encompass greater diversity is a mixed blessing for the UN. Different cultural and disciplinary perspectives certainly enhance debates. However, reconciling widely disparate beliefs and values adds to the challenges that the UN faces in reaching agreement on vital issues, such as deciding upon the Principles for Water that should be applied. There are clearly some gaps between the ideas and ways of thinking that have tended to dominate proceedings thus far, and the views of the more diverse subaltern groups that have recently entered the conversation.

Rather than foregrounding the implicit (and sometimes explicit) ideological and political conflicts inherent in this equation, the UN has framed these differences in terms of values. This is not an isolated effort: there is increasing recognition that values are central to debates about water. For example, the Vatican recently raised the issue of valuing water in broader terms as part of its mission (Tomasi 2017), and it is becoming clear that, while technical obstacles are real, it is the choices made about these that are critical. As Groenfeldt puts it:

> Somehow we have gotten used to the idea that water management is a technical subject better left to the experts. . . Water management is technical, but there are lots of value assumptions embedded in the technical choices. Moreover, the governance of water, the laws, policies, and institutions that set the context for water management, is anything but technical. Water governance is all about values. (2013, 3)

An additional complexity for the UN (and in policy-making generally) is the diversity of views about what values are. In discourses relating to water, there is a considerable gulf between narrow technical or economic concepts of value, generally seen as a form of quantitative measurement, and more complex and largely qualitative social concepts of value. Cultural groups can have very different approaches to value: Indigenous communities in particular rarely conform to dominant notions that values are either "cultural" or "natural" (Agnoletti and Santoro 2015). They also tend to have more holistic concepts of value, integrated with multiple domains of their lives (Harmsworth, Awatere, and Robb 2016; Strang 1997). As Graeber observes, a comparison of diverse ways of approaching value and meaning-making can therefore be revolutionary, disrupting more reductive economist paradigms (2001). It is in bridging these gaps, between different ideas about value, that the theoretical frameworks and the cultural translation provided by the social sciences and the humanities is vital.

CREATING NEW PRINCIPLES FOR WATER

With funding and administration from the World Bank, and led by the Netherlands, the UN's High Level Panel on Water commissioned a team of experts to write background papers that, by elucidating ideas about value, would provide the basis for widespread discussions between representatives from its member states and from major international water-related organizations. From this discussion they hoped to gain agreement on a formal set of principles. My small contribution to this very large conversation was, first, to respond to a request for a foundational paper on cultural and spiritual values relating to water and then, as it turned out, to assist the panel in discussing and writing the first draft of the Preamble and Principles.

As I discovered in working with UNESCO's water-related programs for a number of years, bringing social theory into this kind of forum contains significant challenges for scholars from the social sciences and humanities. This was particularly the case with the Ecohydrology Programme, whose Steering Group (until I was invited to join it) was entirely composed of hydrologists, engineers, and ecologists. Their preference, rather than engaging with social theory, was to employ what they called social hydrology: in essence, social science "reinvented" by engineers. Bringing real social theory into this context was uphill, but eventually a third element of social science was established, at least for as long as the Steering and Advisory Groups continued to include social scientists. But, as this implies, the ongoing participation of social science and humanities scholars is critical in such endeavors.

As Bille Larsen observes, ideas about culture are now broadening (2017), as are understandings of cultural heritage (Brumann 2017). However, the UN continues to divide "values" into three areas—"economic," "environmental," and "cultural and spiritual"—and to assume that these are somehow separate and not equally reflective of "cultural" beliefs and values. This is, of course, a reflection of dominant values in itself. The High Level Panel for Water team leaders accepted my point that all values, including those relating to economic and environmental issues, are intrinsically cultural, but observed that we would still have to work with—or, better still, try to shift—popular understandings of "culture" as some kind of separate domain.

Nevertheless, there are some advantages to the UN's conventional insistence on different categories of value, as they highlight the reality that short-term economic concerns tend to override more complex cultural beliefs and values that are focused on long-term sustainability. In producing a background paper for them, I therefore underlined the point that cultural and spiritual values permeate all human engagements with water, including those focused on economic or ecological concerns, as these too are formed by particular beliefs and understandings. When we describe values as being "cultural" or "spiritual" in nature, we generally mean the deeper, more complex values that hold societies together and that are sometimes subsumed by immediate pressures to meet material and economic needs.

Thus, there is some useful potential to discuss "cultural and spiritual" values as a way of critiquing unsustainable political and economic short-termism. This is urgently needed, as it is plain that most water policies and practices are focused heavily on responding to pressing exigencies, such as the provision of clean water and sanitation or the generation of hydroelectricity to enable economic growth. In doing so, while they may meet immediate social needs, such policies often externalize the costs to less powerful human communities, to future generations, and to nonhuman species. The dominance of short-term economic priorities is readily visible at a local level: the task for the UN is to make this dominance more visible at national and international levels, to the extent that questioning short-term economic priorities becomes part of international debates and promotes the reform of national water policies. Key to driving these changes is the equal consideration of "cultural values" focused on the need to protect ecosystems, to uphold the rights and interests of nonhuman species, and to ensure their well-being (as well as that of future human generations).

A first step, therefore, was to highlight the reality that some "values" are given more weight than others, to underline their relationality, and to provide some definitional clarity. This entailed challenging centrally positioned scientific views that *value* means something quantitative. I therefore underlined to the panel that values are a way of articulating what we think is important and should be prioritized and protected. They compose individual and collective identities and define relationships between human groups and with the nonhuman world. While formal valuations tend to focus on reductive quantitative measures and may be material (e.g., pH measures) or monetary, these sit within a broader context of personal and societal values that are less readily condensed.

Understanding values therefore rests on three key points. First, we need to recognize that all values are relational, involving comparative judgments about what matters the most. Water management reflects decisions about priorities. It is also an exercise of power in which it is critical to consider whose values are being applied and whose interests (human and nonhuman) are—or are not—being met. Second, all values are ingrained: people are habituated into values through education, everyday practices and rituals, social approval or disapproval, and the representation of values in various media, as well as via social and material norms. Specialized contexts (such as particular industries) inculcate subcultural values. Regular engagement with a value-laden context literally inscribes values into the brain, wearing "pathways" to particular patterns of thinking. Third, while quantitative valuation methods seem to offer informed and transparent decisions, many of the most important "cultural" values are difficult and sometimes impossible to quantify. It is here that the social sciences and humanities can really broaden the discussion. Value assessment increasingly includes at least superficial qualitative data, but it remains challenging to articulate complex values, such as what heritage means to people, the aesthetic and spiritual experiences that water enables, or the many and diverse contributions to human and nonhuman well-being that ecosystems provide. To include complex values fully and transparently in water management and use, there is a need to make better use of qualitative evidence (such as that produced by ethnographic and historical research) and to employ analytic approaches from the social sciences and humanities.

Understanding that values are relational highlights the inherent tensions within concepts of sustainable development as they apply to water and to all material engagements. Illich has argued that the concept of sustainable development—as it is currently understood—is fundamentally an oxymoron, since conventional visions of development assume continual economic growth and expansion (1999). Thus, values are not merely relative, they can be irreconcilable. The UN Sustainable Development Goals exemplify this problem. Goal 6: "Ensure access to water and sanitation for all." Goal 7: "Ensure access to affordable, reliable, sustainable and modern energy for all." Goal 13: "Take urgent actions to combat climate change and its impacts." Goal 14: "Conserve and sustainably use the oceans, seas and marine resources." These worthy goals rest on critical values about health and social and ecological justice, but even with greater efficiencies they are unlikely to be met as long as concepts of growth and development remain wedded to each other. Unless artificially increased by unsustainable uses of energy and resources, freshwater resources are finite: all water uses represent value choices that are ultimately sustainable . . . or not.

As illustrated by the photograph of the salination created by producing profitable but soil-damaging cotton in arid regions (figure 9.1), it is inevitably easier to prioritize short-term gains over long-term losses, particularly if the latter accrue to other groups, future generations, or nonhuman beings. But such short-termism is literally costing the Earth. Water is the perfect mirror of relations of power within and between societies and, I would suggest, equally reflects relations between human groups and nonhuman and material worlds. From a local to global scale, inequalities in human and nonhuman relations are manifested in all of the everyday choices made about water management and use. Societal values are writ large, in particular, in choices about infrastructure, which manifest priorities on a grand scale and over long periods of time (Strang 2020).

Locally, nationally and globally, just as the interests of less powerful human communities are persistently overridden, so too are the needs and interests of nonhuman species. In the last few centuries, and most particularly in the last century, human population expansion, along with ever-increasing levels of habitat destruction and water redirection to meet human interests, has led not only to major displacements of human communities, but also to a massive—and still rapidly accelerating—spike in species extinctions, increasing the "normal" rates of

FIGURE 9.1. Salinated landscape in Uzbekistan, where irrigation to produce cotton has raised salts to the surface, rendering the soil infertile. Photo courtesy Veronica Strang.

species extinction by about 10,000 percent and, in the last forty years, reducing the populations of nonhuman species by 60 percent (WWF 2018).

Meanwhile, as noted previously, 40 percent of the world's population is affected by water scarcity, and the UN anticipates a much larger shortfall between supply and demand by 2030 (United Nations High Level Panel on Water 2018, 7). Rising populations will need (on top of the current 70 percent used) 15 percent more freshwater for irrigation and still more to provide sufficient energy supplies (World Bank n.d.). Such demands will push governments to continue to redirect water into agriculture and other forms of economic production, and to prioritize the interests of the most powerful groups. In this economically focused, growth-oriented equation, it is easy to see why primary producers around the world are under constant pressure to expand and intensify their activities further. A farmer might aspire to be a local "guardian of the land" but, under external economic pressures, may sacrifice water quality by using fertilizers to intensify crop production. As Australian farmers often say, "it is hard to be green when you're in the red."

UNITING NATIONS AROUND WATER

How can we hope to achieve more genuinely sustainable human engagements with water in which "development" comes to be about doing it better rather than doing it more? How do we achieve the international and intercultural cooperation needed to solve the problems created by untrammeled growth in human numbers and in human activities?

The United Nations' primary role is to enable inclusive and constructive conversations between societies. Such international bodies are inevitably cumbersome, and deeply hampered by myriad political agendas and people pulling in different directions. But global cooperation depends absolutely on human societies having just such conversations, and, as I noted at the outset, we are now living, uniquely, in an age when it is technologically feasible to do so in real time. It is now possible to go to The Hague to discuss water issues with experts from around the world, and to have a much wider tranche of them participate via virtual means in the conversation. It is possible to write in collaboration with distant colleagues and to circulate material for immediate discussion with high numbers of international participants. So we have the technology; the question is, do we have sufficient will and patience to bring local perspectives into a larger conversational forum and to persist with negotiations that will lead to real change?

Although UN networks constitute a community of sorts, working at this abstract international level is a very different kind of experience for social scientists and humanities scholars. For anthropologists, it raises some challenging questions about how to translate local, ethnographically based understandings of human behavior into highly generalized discussions and takes us into key debates about the comparative nature of our discipline and the extent to which ethnographic findings can be scaled up into meta-discourses to discuss broader human questions. Similar questions can be raised in relation to historical or literary comparison. In fact, in dealing with water issues, this transition is not as difficult as it may seem. This is partly because water itself provides a medium through which it is not only possible but necessary to think across cultural and historical boundaries. Its material properties and behaviors carry one's thinking up (and down) through the various scales and systems through which water flows. In this sense, "thinking with water" provides a useful opportunity to consider how all disciplinary areas might address issues of scale.

In seeking to understand people's relationships with water, I have spent many years, in various parts of the world, meandering up and down river catchments with Indigenous leaders, farmers, industrial water users, recreational and domestic water users, and the public and private companies responsible for managing water flows and treating and delivering water. In accord with their particular cultural and geographic contexts and their specific activities, each of these groups has its own unique ways of thinking about, engaging with, and valuing water. Their different engagements with water often lead to conflicts that are intrinsically about values.

For example, in countries such as Australia and the United States, Indigenous communities, local catchment groups, and recreational water users often express deep anxieties about the effects of mining activities on water quality. In North Queensland, where gold extraction relies on the use of cyanide, riparian water users are concerned about how such poisons leach from storage dams into local

FIGURE 9.2. Tailings dam at Red Dome gold mine, Queensland. Photo courtesy Veronica Strang.

water courses (figure 9.2). There is also rising conflict about the effects of overabstraction on local ecosystems, and downstream farmers rail against the effects of the redirection of water by upstream irrigators.

But, because water flows through everything, such local conflicts are replicated at regional, national, and international levels. Social justice organizations protest against the enclosure and privatization of water, and political factions pull in different directions on issues of ownership and control. Conservation organizations criticize national and international policies that fail to protect the environment. Tiny hydro-squabbles about abstraction from local wells and aquifers, or over the direction of small streams, are echoed in major transboundary river conflicts.

Water is therefore very useful for thinking upwards—or enabling us, as Nader suggested, to "study up," both in terms of understanding wider contexts of governance and power and in thinking about the larger-scale social and ecological dynamics in which local water engagements are nested. All of the differences in values relating to water that emerge at a local level flow upwards into larger-scale debates about water policy. How should water be governed, and by whom? Which—and whose—interests should be prioritized?

In working with the UN, the key challenge was to scale up—to articulate cultural and spiritual values in relation to water in a way that would be meaningful not only at different scales, but also in highly diverse cultural contexts. Fortunately, water is so central an element in every domain of human life that, while all engagements with water are unique to their particular cultural and subcultural contexts, water also holds powerful cross-cultural meanings that recur in each of these, albeit manifested in specifically local forms. It is an area in which it is unusually easy to bridge different cultural perspectives. With a view to supporting a shared conversation about values, I therefore drew on earlier ethnographic research to pull out some core themes of meaning around which important cultural values about water typically coalesce.

Water as Life

The overriding value in relation to water recognizes its essential role in hydrating and thus supporting life and health in all biological organisms. Landscapes are enlivened by water, and no ecosystem can function without it. The concept of water as a source of life is central to creation stories—for example, by Christianity's Genesis, in which God forms the world out of watery chaos; by the emergence of the Mayan world from primal seas on the back of the great water deity Itzam Na; and by the Aboriginal Rainbow Serpent in Australia—an ancestral personification of water from which all life emerges. It can be seen, for that matter, in a contemporary scientific view that, having simmered in the oceans' depths for nearly four billion years, life on earth emerged from the water and "became" the species of today.

These foundational meanings and values continue to flow into everyday engagements with and conflicts over water, such as the Standing Rock controversy in the United States, where the Dakota Sioux, aided by a range of like-minded groups, have stressed that "water is life." A universal understanding that all living kinds—human and nonhuman—depend on water underpins long-standing ideas about water as a common good, and it places it centrally in debates about human and nonhuman rights, and social and ecological justice (Strang 2016a).

Water and Connection

At the most basic molecular level, water molecules have a particular capacity to bind with and thus carry other materials. Where this is most important in terms of meaning and value is that it is plain that water flows through and connects all biological organisms with each other. As McMenamin and McMenamin have pointed out, when terrestrial species crawled out of the oceans, they brought the sea with them, retaining a common organic dependence upon salt and water which the McMenamins describe as a "hypersea" (1994). And the flow of water through plants and animals links them to the material world, composing hydrologically and conceptually linked systems.

Like other aspects of water, this connectivity holds true at every scale. Water moves from cell to cell, irrigating organisms; soils, plants, and animals necessarily ingest or absorb water, integrating it into physical systems; groundwaters form invisible links beneath places; rivers join communities along their courses in a series of upstream and downstream relations; water supply systems connect houses and businesses within cities; and water infrastructures move water between rural and urban areas. At a planetary scale, oceans serve both to separate and connect continents. And people are well aware that they live in concert with planetary streams of hydrological movements.

Cultural values about the centrality of water as a connective substance emerge in multiple domains. We can see them in religious practices in which water is used ritually to symbolize social and spiritual connection. They surface in discourses about the meaning and value of blood connections. Blood may be "thicker than water," but it is largely composed of it, enabling many important ideas about fluid "ties" between kin and intergenerational flows of identity. Fluid interconnections also feature strongly in scientific narratives. For example, the concept of "an ecosystem" is fundamentally a vision of material flows connected by water. At a planetary scale, Vernadsky's notion of the "biosphere" conceptually and materially links all living kinds.

Water and Spiritual Being

As the creative "substance of life," water is often seen as an analogue of human spiritual and social identity, carrying people into and out of corporeal being. This coheres with beliefs that life entails "becoming" material—i.e., embodied— and sentient, while death brings a loss of material form, consciousness, and memory. Just as primal waters generate whole worlds in stories of cosmogenesis, water is often seen as a creative well for human lives and as the medium through which persons are returned to the fluidity of nonbeing. In Classical underworlds, therefore, the rivers Styx and Lethe involve departing and "forgetting."

In some cultural contexts, human and nonhuman life cycles are tied conceptually to hydrological cycles: to visions of life emerging from waters within the land. In Australia, for example, the Rainbow Serpent illustrates just such a hydro-theological cycle. Held within water places in the landscape, it generates human spirits, which then "jump up" to enervate the fetus in a woman's womb. This also illustrates a powerful idea about belonging: a belief that the person is "made of" the substance of that place and retains an inalienable social and material connection with it. As well as locating individuals spatially, this defines their place in a network of kin and their collective rights to clan estates. Upon death, the person's spirit must be ritually "sung back" to its place of origin, to be reunited with (dissolved back into) the Rainbow Serpent being. Such powerful beliefs underpin contemporary ideas about identity and cultural heritage, and understandings that societies are intimately— literally substantially—connected to the places they inhabit.

As well as generating human beings, the Rainbow Serpent is venerated as the source of all living kinds. Similar ideas are evident in the "rain shrines" found in many areas of Africa, which celebrate rainbow serpent beings connecting creatively with earth deities. They are visible in the rain-making rituals of native peoples in America and Canada and in their adherence to the notion of sacred, generative landscapes that has given such impetus to the recent protests about the Dakota pipeline.

Such meanings and values also explain why many cultural traditions contain ideas about "living water" or "holy water": water that contains spiritual essence or which has the power to bless, to cleanse and heal, to enlighten, and to enable transitions through life (and death). Concepts of spiritual and social belonging are intertwined, and water rituals express community, whether through the baptism of strangers, so that they will be recognized by sentient nature beings in Aboriginal cultural landscapes, or through the baptism that incorporates newcomers into Christian congregations.

A related idea imagines water as a source of enlightenment and spiritual wisdom. Biblical and Qur'anic literature provides plentiful images of water as a stream of knowledge or *fons sapientae*, and there are many historical and contemporary cultural traditions in which people seek wisdom from wise ancestral water beings via rituals of engagement with water. Returning to Australia, for example, the Rainbow Serpent is seen as the source of all knowledge, and—in Cape York— secret sacred knowledge is gained through a ritual of immersion, described as "passing through the rainbow."

In a world where most people continue to hold religious beliefs of one kind or another, such ideas still exert a powerful influence on cultural values relating to water. Beliefs about water as the substance of the spirit affirm ideas about water as public good, strengthen views on ownership and privatization, and create extreme responses to issues affecting water quality.

Beliefs about water and the spirit have also segued, unproblematically, into secular ideas about water as a focus for enlightening meditation and experiences of belonging and "oneness" (or in Confucian terms "harmony") with nature. This connects with recreational water use and people's sensory and aesthetic experiences of engaging with water: the fun of playing with water, the restfulness of immersion and weightlessness, or the pleasure of gazing at water. These are all powerful sensory and affective experiences, and therefore generate real emotions—which, as well as creating demands for access to waterways, reservoirs, etc., also flow into values relating to conservation, anxieties about pollution, and so forth.

Water and Agency

No form of production is possible without water. It is fundamental to human capacities to act upon the world, to make things, to grow crops and maintain livestock— in other words, to gain social and economic capital. Water is equally vital to the reproductive capacities of nonhuman species and ecosystems. Human instrumentality in managing water invariably involves choices about which groups' and/or which species' generative capacities are most valuable. This highlights the most critical issue in water management: the tension between human desires to redirect water flows to support their own productive activities and a growing recognition that practices depriving ecosystems of sufficient flow to reproduce themselves are unsustainable and lacking in ecological justice.

Water also has its own agentive powers—a major thread in this book—and its actions upon the material world are not confined to hydrating biological life forms. Oceans stabilize planetary systems. The movements of water shape landscapes, carving valleys and lakes and forming wetland areas. Water flows irrigate, carry fertilizing sediments, and remove waste matter. Water enables the movement of people and goods, and of aquatic species. Its physical forces have long been harnessed to produce energy, via ancient water mills and now through hydropower. The potential of water to produce "work" has long been part of the active process through which societies reproduce themselves.

The value of water therefore encompasses a range of direct and indirect agentive capacities. This reality flows into debates about irrigation and issues about overabstraction and salination, into upstream-downstream conflicts, into disagreements about water charges and allocations, into dissent about environmental flows, and into concerns about controlling floods and droughts.

Water as Order

Water flows literally sustain social, economic, and political order. The movements of water also provide a compelling metaphor about other "flows": economic flows involving "circulations" of resources, "out-of-control" floods or droughts in the market, and notions of "trickling down." It provides a way of thinking about intergenerational flows of knowledge and belief, as well as the flooding, leaking, or seeping of ideas and information from one group to another. It is used to manifest ideas about identity and the circulation of people within and between groups and between geographic regions.

Thus, each of the various meanings of water contains a notion of "orderly" (right) and "disorderly" (wrong) flows, expressed through powerful ideas about purity and pollution. Concepts of pollution are universal, being concerned with "matter out of place." Out-of-control flows of water provide a metaphor for disorder, giving a powerful emotive dimension to floods that contaminate domestic spaces with "foul water" or to anxieties about mine leachate. Drowning metaphors articulate ideas about being "overwhelmed" by emotion. They are employed to describe foreigners "swamping" local social identities or the "pollution" of unwelcome ideas. Images of drought describe not only failing crops and dying cattle, but also markets "drying up," emotional deserts, and the "dust to dust" desiccation of aging.

These metaphorical images of water contain substantial symbolic meaning, and this leads to quite different ideas about what is deemed to be orderly. For environmental activists, order generally means maintaining sufficient ecological flows, while for business interests order may mean ensuring more redirection of water into productive activities. Anxieties about order are also evident in responses to disorderly flows, such as floods and droughts, and may be manifested in demands for flood defenses, water storage, more water security, desalination plants, and such. And fears about the disorder of pollution flow into debates about water quality regulation, environmental legislation, and penalties for polluters.

Water as Health

Long-standing recognition about the importance of water in relation to human health is illustrated by the earliest recorded libations, intended to "revive" (i.e., rehydrate) the mummified bodies of Egyptian pharaohs. They can be seen in the votive offerings by Celtic tribes and Romans to the water deities of healing springs and wells and in the historical reformation of these sites as "holy wells," expressing the miraculous powers of saints and prophets. Zamzam, a sacred well near the Ka'ba in Mecca, has been the focus of pilgrimages for millennia, and ninth-century scholar Ibn al-Faqih noted that its waters provide "a remedy for anyone who suffers."

In a more secular age, many European wells became spas offering "healing waters" and providing careful scientific lists of their health-enhancing minerals. Today, ideas about water as a source of health and vitality are readily visible in the marketing of bottled spring water and in the images of babies gamboling underwater or spouting geysers that persuade people to pay well over the odds for such water's "reviving" qualities. This also costs the earth, as thirteen liters of freshwater are required just to make the bottle.

So, it is clear that, despite changes in form, the notion of "living" health-giving water has persisted tenaciously across time and across cultures. Recognition of water's centrality in maintaining human and nonhuman health is evident in efforts to provide all human communities with clean water, for example in the UN's goal "to ensure access to safe water sources and sanitation." It surfaces in medical advice about drinking sufficient water, and in regulations designed to protect water quality both for drinking water and to sustain the health of ecosystems. It is foundational to debates about water and environmental management.

Water as Wealth

In English, "wealth" and "health" are etymologically and conceptually related, both words connoting "wellness" as being "hale" or "whole." Water enables individual bodies, families, communities, and nations to maintain social and physical integrity and well-being. To hold water—with a dam, in a lake, through owning water allocations or shares in a water company—is to hold wealth. This is readily demonstrated in the high status of water features: spurting generative fountains that grace national monuments, parks, mansions, and private gardens.

The centrality of water to all forms of wealth creation is starkly illustrated in images of wealth and poverty. The rich inhabit lush green spaces, have hot and cold running water, and wealth signifiers such as swimming pools and waterfront homes. The poor are marginalized in deserts or gardenless slums: without proper or clean water supplies, they must trek for miles toting water containers or take their washing down to the river. Imagined paradises and utopias are well-watered, green, and fertile. Hell involves fetid slime, blasted deserts, or fire.

The underlying meanings of water as wealth provide a broader context for the specifically economic valuations of water that generally dominate discourses relating to water management and use. What may seem like entirely pragmatic measurements of water allocations, crop yields, and contribution to GDP are nested within much more complex cultural values about individual and collective abilities to harness water's generative power and to be socially and economically productive. Notions of "wealthiness" are collectively expressed in notions of the common good or "the Commonwealth." The conflation of water and wealth lends intensity to debates about ownership, control, and access to water, and these debates provide a context for conflicts over governance, legislation, water supply, charges for water, and choices about priorities for water use.

Water is also central to conflicts about "growing the economy." The "growth is good" mantra obscures the reality that this involves important value choices. The growth of some things, or the promotion of particular interests, may be at the expense of others. Water redirected into hydropower may benefit national economies but prevent the flourishing of local communities or ecosystems. Irrigation schemes—as illustrated in areas such as the Murray Darling Basin or the Colorado River—may promote the growth of "high value" crops such as cotton or rice, but do so at the expense of native vegetation and biodiversity.

Water as Power

The control of water is fundamental to political power and to the sovereignty of nations, and arrangements around the ownership and control of water directly reflect the realities of power relations within societies. This has been most famously illustrated by Wittfogel's account of the relationship between the development of vast canal systems in China and the power of imperial dynasties (1957). But it is also readily evident in India, South Korea, and many other parts of the world, where major dams and irrigation schemes remain entangled with aspirations for nationhood, and with the power of social and political institutions. This also applies within states: California continues to experience internal "water wars," and in Australia there are some lively upstream and downstream controversies between Queensland and NSW about flows across the border into the Murray Darling Basin (figure 9.3).

The notorious case of Cubbie Station, which captures about a quarter of the water that would otherwise flow across the Queensland border into the Darling River, also reminds us that, in a global economy, water privatization has begun to shift control from governments to transnational corporations. This important trend gives increasing control over freshwater flows to institutions that are socially and physically detached from local communities and their environments. Such "disembedding," as Polanyi called it (1957), raises major issues around ownership

FIGURE 9.3. In 2019 over a million fish died in the Murray-Darling Basin because of the algal blooms and loss of oxygen caused by insufficient water flows. Photo Wikimedia Commons.

and about democracy itself. As I have noted elsewhere (Strang 2016b), governments can only govern effectively to the extent that they can control essential resources and ensure water security. Social scientists should be asking pointed questions about whether and how governments can do so when land and water are owned by transnational corporations with no local accountability.

Forms of water ownership range from collective (common property regimes or national ownership) to privatized water supply companies and water markets. Attempts to govern the latter are expressed by regulatory regimes aiming to protect the rights and interests of people and environments, but there are major variations in regulatory capacities to meet these aims, and they are rarely adequate to the task.

Each form of ownership and regulation empowers or disempowers particular groups and expresses particular values. This connects directly with the extent to which water is regarded as a common good. Founded on moral questions about rights to the substance of life, access, and equity, the concept of water as a common good is central to historical and contemporary conflicts over water, whether at an international or local scale. It appears in historical accounts of negotiations about water releases between mills along medieval waterways, just as it does in contemporary discussions about transboundary flows on the River Jordan, or in every hydro-squabble about water allocations.

Power via the ability to direct water is not confined to governments, NGOs, and corporations or even to large-scale institutions. Scientists and conservation organizations participate in decisions about waterways through the application of their expertise. Community river management groups act directly to protect the health of their local waterways. Farmers' direction of water flows enables them to produce life's necessities and sustain their own and wider communities, and this role in feeding the world, in managing the land, is intrinsic to their subcultural identity, just as washing ore to produce gold or using water to make industrial products underpins the identity of miners and manufacturers. Such positive values are echoed by water engineers in the United Kingdom, who recall the heady days before privatization of being the "heroes" empowered to deliver water to the domestic tap. But it is also essential to note that there are major asymmetries in power relations too, and that some groups—such as Indigenous communities, ethnic minorities, women, and the poor—continue to struggle to participate in decisions about water, within nations and internationally.

How water is controlled and distributed is therefore, always, a direct reflection of social and political relationships, not only between humans but also between humans and other species. In this sense, decisions about water flows—and about whose human/nonhuman needs and interests are met—are immediately expressive of societal values about the environment.

Water and Cultural Heritage

The UN is paying increasing attention to the interconnections between water and cultural heritage. It is clear that water flows alongside both tangible and intangible cultural heritage. It is central to all societies' processes of production and reproduction—social, economic, political, and spiritual—and to the dynamic composition of cultural land and waterscapes. While there are thousands of important "world heritage" sites, these are not merely remnants of historical activities: they are a living part of cultural engagement with the material environment that reproduces particular lifeways. It is for this reason that in composing a Water Framework Directive in 2000, the first point articulated by the European Union was that "water is not a commercial product like any other but, rather a heritage which must be protected, defended and treated as such."

It is relatively easy to see why successive generations of farmers develop attachments to the land they have irrigated, and to the soils they have nurtured and made productive. In Australia, the outback "battlers" who have built up their farms over generations are central to Australian history and cultural heritage, and one might find similar narratives in many other histories of settlement. Yet in industrial societies farms are often alienable properties, and it is not unusual for farmers to sell up and move on. Still, nations continue to valorize this cultural history, and this is readily evident in political debates and in art, literature, and other media.

Cultural heritage is central for any group, but it is perhaps particularly important for communities who constitute ethnic or cultural minorities and who are trying to maintain traditional lifeways in larger societies. Cultural heritage carries quite different meanings and values for long-term place-based societies, such as Indigenous Australians and First Nation peoples, whose activities are mediated by permanent relationships with sentient land and waterscapes. In such contexts, cultural and spiritual values relating to water run deeper. To understand these close affective relationships between people and places, we therefore need to think more deeply and be willing to learn from very different cultural perspectives.

The centrality of water in relation to cultural heritage is apparent in every debate about how water should be owned, used, and controlled. It is readily visible in the concerns of small rural communities facing major upstream irrigation corporations. It is intrinsic to every land claim by Indigenous communities. It can be seen in the responses that are generated by every proposed development scheme.

Articulating Cultural Values

Another question the UN asked me to address was: how can values relating to water be articulated? Methodologically, it is possible to skate over the surface with surveys, for example by ranking people's concerns about waterways or their "willingness to pay" (WTP) for various measures, but deeper cultural values are not readily quantified. Social scientists and humanities scholars bring to the table robust qualitative methods that can make more complex values visible and comprehensible. Making use of archaeological and historical research allows us to compare relationships with water over time; and ethnographic research methods from anthropology (interviews and long-term participant observation) can create a detailed picture of diverse beliefs and values relating to water, and the everyday practices through which these are expressed and maintained.

As well as examining everyday practices, it is also useful to consider the use of water in religious and secular rituals. A nice example of the latter is provided by the *Splash!* Festival in Queensland, in which communities bring vessels of water from local waterways and pour these into a shared vessel to articulate how local waterways connect them to each other (figure 9.4). No major dam or irrigation scheme opens without rituals celebrating its capacity to control the power of water, and no spouting fountain, displaying the *mana* of a town or state or nation, is switched on without public ceremony and the participation of suitable dignitaries. These things are not difficult for social scientists to explicate.

However, the use of qualitative methods raises several related questions. What kinds of outputs are needed to communicate complex cultural values, and how can we ensure that these are incorporated into decision-making processes and given sufficient weight? Decision-making processes are currently heavily reliant upon economic models and reductive methods of presenting information in largely quantitative form. This is also indicative of the dominance of the values promoted by economic actors: keeping discussions focused on quantitative measures maintains

FIGURE 9.4. Splash! Festival water ritual in Maroochydore, Queensland. Photo courtesy
Veronica Strang.

the primacy of their concerns. Introducing more qualitative material therefore
meets resistance, not only from busy policy makers keen to have highly condensed
information, but also from those for whom a reliance upon reductive discursive
forms serves to exclude issues that might conflict with their particular interests.

Will giving more equality—and higher visibility—to cultural values mean
that deeper concerns are consistently and equally reflected in water management
decisions? It is not useful to valorize clean water for ecological or moral reasons
on the one hand, if permission is given to money-saving polluting activities on
the other. Different government agencies may be said to promote particular—
economic, environmental, and cultural—domains of value. Their relative influence
and resourcing illustrates the point made at the beginning of this chapter: values
are relational, and debates are fundamentally about which—and whose—
values matter the most. Realpolitik ensures that priority is generally given to
short-term exigencies: the meeting of basic needs and the promotion of economic
growth. The extent to which such priorities are ameliorated by longer-term thinking depends upon the political climate. Thus, when über right-wing politicians
and climate change deniers have a free hand, it is more or less inevitable that funding cuts will be imposed upon environmental agencies or that environmental regulations will be relaxed.

As this implies, what differentiates key values the most is whether they focus on immediate or long-term interests, on the interests of a few people or of all, and whether they encompass nonhuman needs and interests. Good leadership deals with exigencies but also considers the wider view and the longer term. It encompasses the well-being of all living kinds, both for moral reasons and because it recognizes the complex interdependencies that connect humans, nonhumans, and the material world. In this sense, the "common good" of water represents the wider common good of collective sustainability.

In my work with the UN, I have therefore tried to promote the idea that cultural and spiritual values are the deeper values of societies. Such values look beyond short-term social and economic needs to consider future generations and future worlds. They support more conservative use of resources and higher levels of protection for less powerful human and nonhuman populations. They have the potential to counterbalance values focused on short-term economic gains and to question assumptions about "growth." Making cultural and spiritual values explicit, giving them sufficient weight, and including them fully in debates about water use and management is therefore a matter of urgency. They provide the holistic, long-term view that is needed to shift decision-making towards more sustainable water policies and practices. I proposed several key principles that might be adopted:

Principle 1. Recognize and respect cultural beliefs and values and cultural heritage relating to water.
Principle 2. Fully include and give equity to diverse cultural and spiritual values, and their proponents, in debates and decisions relating to water policy and practice.
Principle 3. Prioritize cultural values and practices that promote social and ecological justice and protect the health of all human and nonhuman beings.

Having submitted the background paper and participated in a workshop in The Hague, I was invited to join the core writing team and to help draft the Preamble and Principles that would initiate a wider international conversation. The group wrote a strong draft, knowing that, as with all sensitive political matters that require consensus from diverse groups, any contentious points are likely to be diluted to almost homeopathic levels during the consultation process.

Following wider international conversations over the next few months, a set of agreed Principles emerged as part of the High Level Panel's outcome report in March 2018. In the first instance, the report reaffirmed its allegiance to quantitative data, observing:

The adage "you can't manage what you can't measure" is particularly true for water. Information about water quantity, quality, distribution, access, risks, and use is essential for effective decision-making, whether by businesses managing a production process, rural communities managing a well or basin authorities managing a flood. (United Nations High Level Panel on Water 2018, 16)

However, the report also linked the issue of water to the range of UN Sustainable Development Goals and stressed that "Societies need to value the water they have—in all its social, cultural, economic, and environmental dimensions" (2018, 13). It articulated the following principles:

Recognize and Embrace Water's Multiple Values

1. **Identify and take into account the multiple and diverse values of water to different groups and interests in all decisions affecting water.**
   There are deep interconnections between human needs, social and economic well-being, spiritual beliefs, and the viability of ecosystems that need to be considered.

Reconcile Values and Build Trust

2. **Conduct all processes to reconcile values in ways that are equitable, transparent, and inclusive.**
   Trade-offs will be inevitable, especially when water is scarce, and these call for sharing benefits amongst all those affected. Inaction may also have costs that involve steeper trade-offs. These processes need to be adaptive in the face of local and global changes.

Protect the Sources

3. **Value, manage, and protect all sources of water, including watersheds, rivers, aquifers, associated ecosystems, and used water flows for current and future generations.**
   There is growing urgency to protect sources, control and prevent pollution, and address other pressures across multiple scales.

Educate to Empower

4. **Promote education and public awareness about the intrinsic value of water and its essential role in all aspects of life.**
   This will enable broader participation, water-wise decisions and sustainable practices in areas such as spatial planning, development of infrastructure, city management, industrial development, farming, protection of ecosystems and domestic use.

Invest and Innovate

5. **Ensure adequate investment in institutions, infrastructure, information, and innovation to realize the many different benefits derived from water and reduce risks.**
   This requires concerted action and institutional coherence. It should harness new ideas, tools, and solutions while drawing on existing and Indigenous knowledge and practices in ways that nurture the innovative leaders of tomorrow. (2018: 17)

It will be evident from the finalized Principles that there was considerable compromise in language and priorities during the consultation process. Stronger points, for example about subaltern and nonhuman rights, were heavily encoded. However, they were not entirely erased, peeping through in key phrases about encompassing all diverse "interests" and noting the interconnections between human needs and the "viability of ecosystems." There was an emphasis on making values transparent, in ways that are "equitable and inclusive." Indigenous knowledge was specifically mentioned as contributing to future leadership.

It is also notable that the UN 2018 Report on Water focused almost entirely on promoting "Nature Based Solutions" that, with their implicit remit to work *with* the realities of local ecosystems, rather than merely imposing "infrastructural violence" upon them (Rodgers and O'Neill 2012), might be said to incorporate some acknowledgement of nonhuman needs in politically palatable form.

PUTTING PRINCIPLES INTO ACTION

Since the Principles were established, the High Level Panel on Water and its related groups have focused on trying to bring these "top down" ideas to a grassroots level. They are working with an array of small projects, mostly focused on river catchment research. They brought me back into the equation to meet with the researchers involved in these projects and to discuss with them how they might articulate the kinds of cultural values discussed above. Naturally, my major piece of advice has been to work collaboratively with anthropologists and other social scientists and humanities scholars, rather than (as has often happened in river catchment research) asking hydrologists or ecologists to reinvent social science in order to keep things simple. A concurrent aim, therefore, is to encourage scholars from across the academic spectrum to participate in such collaborations.

Key to the success of such activities is the achievement of disciplinary equality so that social scientists and humanities scholars are involved, from the beginning, in the design and implementation of such research and that their findings are fully and equally incorporated into project outputs. This highlights the importance of current debates about facilitating interdisciplinary research, in which it can be seen that the valorization of different disciplines is reflective of the wider values attached to their objects of study. Thus, STEM disciplines promoting technically instrumental and economic solutions tend to have the best access to funding, and to have dominant roles in interdisciplinary research teams, and the social sciences, arts, and humanities are tagged on, often somewhat tokenistically. So, there is important work to do in achieving disciplinary equality in this area.

There is also work to be done in better understanding the multiple professional and political networks that the UN brings together, and how issues are negotiated between them. In a representational system, national or regional interests may be articulated by a politician, a water ecology specialist, an Indigenous leader, or a representative from a socially oriented NGO. The confusion of multivocal discussions sometimes seems arduously messy and cumbersome. However, the fluidity of confusion is also organic and creative. Coherent narratives do emerge and generate agreement about action: in this instance, a greater articulation of deeper cultural values, a commitment to shifting the discussion to a grassroots level where changes can be enacted, and a push towards Nature Based Solutions. Making use of the latter, I tried to promote a parallel idea of "Culture Based Solutions." My hope is that the (now widespread) recognition that water infrastructure and management needs to fit the realities of local ecosystems might be translated into a parallel notion, that the use and management of water also needs to cohere with local cultural realities.

CONCLUSION

How will the UN's initiative to foreground cultural and spiritual values and seek more sustainable engagements with water have a real effect? Thirty years ago, I had an even tinier role working on the Canadian contribution to the Brundtland Report, which set out plans for *Our Common Future* and promoted the idea that the adoption of sustainable values would save the planet. But since then, as we have seen, anthropogenic impacts on the Earth's ecosystems have worsened dramatically, many societies are struggling with very challenging issues around freshwater, and it is clear that we are heading towards more extreme problems in this critical area.

I would like to be as optimistic about these international efforts as I was in the 1980s, but in the face of dominant ideologies determinedly wedded to growth and competition, the new Principles for Water and the thinking and discussing that they will initiate may be too weak, too little, and too late. We are swimming against the tide. But perhaps, along with the many grassroots countermovements urging real changes, along with emergent ideas about degrowth economics and ecological and social justice, and with the added pressure of sheer terror about water security, maybe—just maybe—there is a tipping point for change that can be reached. The leadership provided by UN High Level Panels and their "sherpas" will be key, as will the diversity of expertise brought to bear on whatever problem is being addressed.

The UN is one of the few organizations that can both provide legitimate leadership and facilitate the international cooperation that is needed. The Principles for Water and the other recommendations of the panel provide a common basis for discussion with policy and decision makers, and we must hope that the combination of top-down and grassroots pressures will encourage them to shift towards more sustainable decisions. I take the view that it is vital to support such efforts. There is surely a critical role for social science and humanities scholars in bringing to the fore the deeper cultural values that are needed to turn societies' current patterns of flow in a more sustainable direction.

REFERENCES

Agnoletti, Mauro, and Antonio Santoro. 2015. "Cultural Values and Sustainable Forest Management: The Case of Europe." *Journal of Forest Research*, 20 (5): 438–44.

Allen, Stewart. 2018. *An Ethnography of NGO Practice in India: Utopias of Development*. Manchester: Manchester University Press.

Arce, Alberto, and Norman Long, eds. 1999. *Anthropology, Development, and Modernities: Exploring Discourses, Counter-Tendencies, and Violence*. London: Routledge.

Bangstad, Sindre. 2017. *Anthropology of Our Times: An Edited Anthology in Public Anthropology*. New York: Palgrave Macmillan.

Berriane, Yasmine. 2017. "Development and Countermovements: Reflections on the Conflicts Arising from the Commodification of Collective Land in Morocco." In *Development as a Battlefield*, edited by Béatrice Hibou and Irene Bono, International Development Policy Series, no.8, 247–67. Leiden: Brill-Nijhoff.

Bille Larsen, Peter, ed. 2017. *World Heritage and Human Rights*. London: Earthscan-Routledge.

Brumann, Christoph. 2014. "Shifting Tides of World-Making in the UNESCO World Heritage Convention: Cosmopolitanisms Colliding." *Ethnic and Racial Studies* 37 (12): 2176–92.

Corsín Jiménez, Alberto. 2007. *The Anthropology of Organisations*. Abingdon, Oxfordshire: Ashgate.

Crewe, Emma, and Richard Axelby. 2012. *Anthropology and Development: Culture, Morality and Politics in a Globalised World*. Cambridge: Cambridge University Press.

Escobar, Arturo. 2005. *Encountering Development: The Making and Unmaking of the Third World*. Princeton: Princeton University Press.

Garsten, Christina, and Anete Nyqvist, eds. 2015. *Organisational Anthropology: Doing Ethnography in and among Complex Organizations*. London: Pluto Press.

Ghali, Boutros. 1992. *An Agenda for Peace: Preventive Diplomacy and Peace Keeping*. New York: United Nations Press.

Graeber, David. 2001. *Towards an Anthropological Theory of Value: The False Coin of Our Own Dreams*. New York: Palgrave Macmillan.

———. 2013. "It Is Value that Brings Universes into Being." *HAU: Journal of Ethnographic Theory* 3 (2): 219–43.

Groenfeldt, David. 2013. *Water Ethics: A Values Approach to Solving the Water Crisis*. Abingdon: Routledge.

Harmsworth, Garth, Shaun Awatere, and Mahuru Robb. 2016. "Indigenous Māori Values and Perspectives to Inform Freshwater Management in Aotearoa–New Zealand." *Ecology and Society* 21 (4): 9. http://dx.doi.org/10.5751/ES-08804-210409.

Herzfeld, Michael. 1992. *The Social Production of Indifference: Exploring the Symbolic Roots of Western Bureaucracy*. Oxford: Berg.

Hobart, Mark, ed. 1993. *An Anthropological Critique of Development: The Growth of Ignorance*. London: Routledge.

Hoedeman, Oliver, Satoko Kishimoto, Martin Pigeon, and David McDonald. 2013. *Remunicipalisation: Bringing Water Back into Public Hands*. Amsterdam: Transnational Institute.

Illich, Ivan. 1999. "The Shadow Our Future Throws." *New Perspectives Quarterly* 16 (2): 14–18.

Johnston, Barbara Rose, Lisa Hiwasaki, Irene J. Klaver, Ameyali Ramos-Castillo, and Veronica Strang, eds. 2012. *Water, Cultural Diversity & Global Environmental Change: Emerging Trends, Sustainable Futures?* Paris: Springer and UNESCO. http://dx.doi.org/10.1007/978-94-007-1774-9.

Kishimoto, Satoko, Emanuele Lobina, and Olivier Petitjean. 2015. *Our Public Water Future: The Global Experience with Remunicipalisation.* Amsterdam: Transnational Institute.

Korten, David C. 1990. *Getting to the 21st Century: Voluntary Action and the Global Agenda.* West Hartford, CT: Kumarian Press.

Lashaw, Amanda, Christian Vannier, and Steven Sampson, eds. 2017. *Cultures of Doing Good: Anthropologists and NGOs.* Tuscaloosa: University of Alabama Press.

Latour, Bruno. 2005. *Reassembling the Social: An Introduction to Actor-Network-Theory.* Oxford: Oxford University Press.

Leve, Lauren, and Karim Lamia. 2001. "Privatizing the State: Ethnography of Development, Transnational Capital, and NGOs." *Political and Legal Anthropology Review* 24 (1): 53–58.

Lewis, David, and David Mosse. 2006. *Development Brokers and Translators: The Ethnography of Aid and Agencies.* Bloomfield, CT: Kumarian Press.

McMenamin, Dianna, and Mark McMenamin. 1994. *Hypersea.* New York: Columbia University Press.

Miller, Daniel, Elisabetta Costa, Nell Haynes, Tom McDonald, Razvan Nicolescu, Jolynna Sinanan, Julioano Spyer, Shriram Venkatraman, and Xinyuan Wang. 2016. *How the World Changed Social Media.* London: UCL Press.

Nader, Laura. 1969. "Up the Anthropologist: Perspectives Gained from Studying Up." In *Reinventing Anthropology*, edited by Dell Hymes, 285–311. New York: Pantheon.

Nash, June. 1979. "Anthropology of the Multinational Corporations." In *New Directions in Political Economy: An Approach from Anthropology*, edited by Madeline Barbara Leons and Frances Rothstein, 173–200. Westport, CT: Greenwood Press.

Polanyi, Karl. 1957. *The Great Transformation.* Boston: Beacon Press.

Robbins, Joel. 2012. "Cultural Values." In *A Companion to Moral Anthropology*, edited by Didier Fassin, 117–32. West Sussex: Wiley-Blackwell.

Rodgers, Dennis, and Bruce O'Neill. 2012. "Infrastructural Violence: Introduction to the Special Issue." *Ethnography* 13 (4): 401–12.

Scott, John. 2000. *Social Network Analysis: A Handbook.* 2nd ed. London: Sage.

Strang, Veronica. 1997. *Uncommon Ground: Cultural Landscapes and Environmental Values.* Oxford: Berg.

———. 2016a. "Justice for All: Inconvenient Truths and Reconciliation in Human–Non-Human Relations." In *Major Works in Anthropology: Environmental Anthropology*, 2:263–78. London: Taylor and Francis. Reprinted from *Routledge International Handbook of Environmental Anthropology*, edited by Helen Kopnina and Eleanor Shoreman-Ouimet, Abingdon: Routledge, 2016.

———. 2016b. "Infrastructural Relations: Water, Political Power and the Rise of a New Despotic Regime." In *Water Alternatives*, special issue: *Water, Infrastructure and Political Rule*. 9 (2): 292–318.

———. 2020. "Materialising the State: The Meaning of Water Infrastructure." In *Shifting States*, edited by Alison Dundon and Richard Vokes, 43–61. London: Bloomsbury Press.

Tomasi, Silvano M. 2017. *The Vatican in the Family of Nations: Diplomatic Actions of the Holy See at the UN and Other International Organizations in Geneva.* Cambridge: Cambridge University Press.

Tsing, Anna. 2005. *Friction: An Ethnography of Global Connection*. Princeton: Princeton University Press.

UNDP (United Nations Development Programme). 2006. "Beyond Scarcity: Power, Poverty and the Global Water Crisis." http://hdr.undp.org/en/content/human-development -report-2006.

United Nations. 2007. "Coping with Water Scarcity: Challenge of the Twenty-First Century." www.un.org/waterforlifedecade/scarcity.shtml.

United Nations High Level Panel on Water. 2018. "Making Every Drop Count: An Agenda for Water Action." HLPW Outcome Report. https://sustainabledevelopment.un.org /HLPWater.

World Bank. N.d. "Water." www.worldbank.org/en/topic/water/overview/.

WWF (World Wildlife Fund). 2018. "Living Planet Report 2018." Washington, DC: WWF.

ACKNOWLEDGMENTS

We coeditors wish to express heartfelt gratitude to all of our friends and colleagues who contributed both directly and indirectly to this book and the years of research and collaboration devoted to its making, with added acknowledgment that its finishing was made immeasurably more complex in the shadow of the global COVID-19 pandemic.

First and foremost, we thank participants in the University of California Merced Water Seminar series (2015–17); the Water and Humanities: The Bridge Is History Conference and Exhibition (2016); the Water: Ways of Knowing, Ways of Being Conference; and the funders who made the two-year postdoctoral fellowship (2015–17) held by De Wolff and Faletti, and directed by Ruth Mostern and Ignacio López-Calvo, a reality. The many faculty, staff, students and public community participants offered ideas, conversation, engagement, and seeds of possibility for the future of water discourse. Partners in these activities included the UC Merced Center for the Humanities, UC Water, the Center for Information Technology Research in the Interest of Society (CITRIS), the UC Merced Library, the UC Merced Yosemite Field Station at Wawona, the Merced Vernal Pools & Grassland Reserve, the Merced County Courthouse Museum, and the International Water History Association. For funding the Open Access publication for this book, we thank the University of North Texas Department of Philosophy & Religion and the UC Merced Center for the Humanities. None of this could have been accomplished without the exquisite organization and dedication of Christina Lux and Austyn Smith Jones; their knowledge of how to get things done was invaluable.

We extend our thanks to Environmental Studies editor Stacy Eisenstark, editorial assistants Robin Manley and Naja Pulliam Collins, copyeditor Ben Alexander, and everyone at University of California Press who marshaled our book project through with enthusiastic support. We appreciate the generous attention given to our manuscript in the review process by our two anonymous reviewers, whose thoughtful critique made the book better.

Faletti and De Wolff are indebted, in the best ways, to Ruth Mostern and Irene Klaver. Ruth led us through the first year of our postdoctoral fellowship, establishing a creative intellectual environment and positive influence that will last a lifetime. Irene, whose wealth of water knowledge and international leadership in water scholarship guided the framing of this book at the beginning, helped to focus the ideas in the introduction, and helped us see the book through to the end in her rare, selfless way. For rich water conversations that have helped develop our ideas, De Wolff is especially grateful to the Oceanic Humanities Symposium at Texas A&M, Lisa Han, Stefan Helmreich, Stephanie Kane, Melanie Yazzie, and Elana Zilberg; and Faletti to her colleagues of the International Water History Association, the Society for the History of Technology, Georgetown University's Mellon Sawyer Anthropocene Seminar and Our World of Water Symposium, the Grasping Waters Institute at the University of Minnesota, to Richard Shiff, and to Maline Werness-Rude.

Last, we extend our pride and our loving thanks to our families for their support of this project, especially during late nights, crunch times, and topographical diversions. We appreciate their patience for the inevitable changes in current that keep moving our work toward, and with, water.

CONTRIBUTOR BIOS

KIM DE WOLFF is assistant professor of philosophy at the University of North Texas, and associate director of the Philosophy of Water Project. Her interdisciplinary humanities research connects global ecological crises to cultures of consumption and waste, to address big environmental questions about everyday life. She has published work on materiality and plastic-water-marine life entanglements, and is currently working on a book project about the "garbage patch" of waste circulating in the North Pacific Ocean.

KALE BANTIGUE FAJARDO is associate professor of Asian American Studies at the University of Minnesota, Twin Cities. He is the author of *Filipino Crosscurrents: Oceanographies of Seafaring, Masculinities, and Globalization* (University of Minnesota Press, 2011, reprinted by the University of the Philippines Press, 2013). He is currently working on a second book titled *Another Archipelago: Diasporic Filipino Masculinities, Place, Water, and Visual Cultures.* He is also working on a coral regeneration project in San Juan, Borikén, Puerto Rico, and is designing a nonprofit organization called Proyecto Kanoa. He lives translocally in Minneapolis, Santa Barbara, Portland, and San Juan.

RINA C. FALETTI is an environmental humanities scholar who investigates histories of water and landscape, art and architecture, and urban modernism, with a special focus on water and wildfire in California and on water imagery in pre-Columbian Mesoamerica and ancient Rome. She teaches histories of art, architecture, and design sustainability at California State University, Chico, and is an affiliated researcher with the Global Arts Studies Program at University of California, Merced. She is the founding curator of Art Responds, a program of exhibitions and public programs responding to environmental crisis and recovery in California. She is secretary of the International Water History Association.

PENELOPE K. HARDY is a historian of science and technology, focusing on technologies of science, ocean sciences, and scientific exploration of the global ocean, and an assistant professor of history at the University of Wisconsin–La Crosse. She has published on topics

245

including military-scientific partnerships in the United States and the United Kingdom, meteorology in interwar Germany, and ocean mapping as both technical feat and imaginative exercise. Her current book project explores the role of research vessels in the creation of oceanography in the nineteenth and twentieth centuries.

STEPHANIE C. KANE is a cultural anthropologist and professor of International Studies at Indiana University Bloomington. Her creative nonfiction on water includes *Where Rivers Meet the Sea: The Political Ecology of Water* (Temple); solo and co-authored articles in *Limn; Human Organization; Anthropocene; Social Text; Political and Legal Anthropology Review; Journal of Folklore Research; Crime, Media and Culture*; and chapters in *Territory beyond Terra; Rivers of the Anthropocene; Transforming Urban Waterfronts*; among others. The working title of her current book project is "Geo-culture: The Ethnography of River City Flood Control."

IRENE J. KLAVER is professor and chair of the Department of Philosophy & Religion at the University of North Texas, and director of the Philosophy of Water Project. She has published and lectured extensively on cultural, sociopolitical, and philosophical dimensions of water. She has coedited various books, including the UNESCO/Springer book *Water, Cultural Diversity & Global Environmental Change*. She likes to dip her toes into visual material water works: she has codirected/produced water films, participated in art exhibits with water photo provocations, and collaborated with architects in *Designing with Water* projects.

IGNACIO LÓPEZ-CALVO is University of California, Merced Presidential Endowed Chair in the Humanities, Director of UC Merced's Center for the Humanities, and professor of Latin American literature. He is the author of more than one hundred articles and book chapters, as well as nine monographs and seventeen edited books on Latin American and U.S. Latino literature and culture. He is the cofounder and co–executive director of the academic journal *Transmodernity: Journal of Peripheral Cultural Production of the Luso-Hispanic World* as well as the co–executive director of the Palgrave Macmillan book series "Historical and Cultural Interconnections between Latin America and Asia" and the Anthem Press book series "Anthem Studies in Latin American Literature and Culture Series."

HUGO ALBERTO LÓPEZ CHAVOLLA is a PhD student in the Interdisciplinary Humanities program at University of California, Merced. He specializes in Latin American studies, focusing on the cultural and literary production of the Arab diaspora in Latin America. His research interests include Latin American literature, Global South cross-cultural relations, and the affective processes in the production of space and culture.

RUTH MOSTERN is associate professor of history and director of the World History Center at the University of Pittsburgh. She is the author of *Yu's Traces: Three Thousand Years of Erosion and Flood Control along China's Yellow River*, forthcoming from Yale University Press, as well as several other articles and special issues of journals on topics in water history and water humanities. She is also the principal investigator of the World Historical Gazetteer project.

ABUBAKR MUHAMMAD is associate professor and chair of electrical engineering and the founding director of the Center for Water Informatics & Technology (WIT) at Lahore University of Management Sciences (LUMS), Pakistan. His interests are at the intersection of environment, technology, and society, covering topics related to hydro-informatics, digital agriculture, and human-water interactions. A humble son of fruit-sellers for many generations, he dreams to leave behind a basin-scale living mosaic of irrigated gardens for his daughters —"*gardens, underneath which rivers flow.*"

CHANDRA MUKERJI is Distinguished Professor Emerita of Communication and Science Studies at the University of California, San Diego, and Chercheuse Correspondante, Institut Marcel Mauss, Paris. She has written extensively on landscapes and water, most notably, *Impossible Engineering* (Princeton University Press), her book on the Canal du Midi that co-won the 2012 Distinguished Book award from the American Sociological Association. She also received the Mary Douglas Prize in 1998 for *Territorial Ambitions and the Gardens of Versailles* (Cambridge University Press), and the Robert K. Merton Award in 1991 for *A Fragile Power* (Princeton University Press).

VERONICA STRANG, Fellow of the Academy of Social Sciences (UK), is an environmental anthropologist and executive director of Durham University's Institute of Advanced Study. Her research focuses on human-environmental relations, in particular, people's engagements with water. She is the author of *The Meaning of Water* (Berg, 2004); *Gardening the World: Agency, Identity, and the Ownership of Water* (Berghahn, 2009); and *Water, Culture and Nature* (Reaktion, University of Chicago Press, 2015).

JAMES L. WESCOAT JR. is Aga Khan Professor Emeritus of Islamic Landscape Architecture and Geography at the Massachusetts Institute of Technology. His research has focused on water systems in South Asia from the garden to river basin scale. He directed the Smithsonian project titled "Garden, City, and Empire: The Historical Geography of Mughal Lahore." At the larger scale he has contributed to studies in the Indus, Colorado, Ganges, Mississippi, Great Lakes, and Amu Darya basins.

INDEX

Founded in 1893,
UNIVERSITY OF CALIFORNIA PRESS
publishes bold, progressive books and journals
on topics in the arts, humanities, social sciences,
and natural sciences—with a focus on social
justice issues—that inspire thought and action
among readers worldwide.

The UC PRESS FOUNDATION
raises funds to uphold the press's vital role
as an independent, nonprofit publisher, and
receives philanthropic support from a wide
range of individuals and institutions—and from
committed readers like you. To learn more, visit
ucpress.edu/supportus.

Milton Keynes UK
Ingram Content Group UK Ltd.
UKHW021958120524
442562UK00008B/42